AF574382

On Invariants and the Theory of Numbers

On Invariants and the Theory of Numbers

Leonard Eugene Dickson

DOVER PUBLICATIONS, INC.
Mineola, New York

DOVER PHOENIX EDITIONS

Bibliographical Note

This Dover edition, first published in 1966 and reprinted in 2004, is an unabridged and unaltered republication of the work originally published by the American Mathematical Society in 1914 as Part I of *The Madison Colloquium* (1913), Volume IV of the Colloquium Lectures.

International Standard Book Number: 0-486-43828-7

Manufactured in the United States of America
Dover Publications, Inc., 31 East 2nd Street, Mineola, N.Y. 11501

CONTENTS

LECTURE III

Invariants of a Modular Group. Formal Invariants and Covariants of Modular Forms. Applications

LECTURE IV

Modular Geometry and Covariantive Theory of a Quadratic Form in m Variables Modulo 2

LECTURE V

A Theory of Plane Cubic Curves with a Real Inflexion Point Valid in Ordinary and in Modular Geometry

On Invariants and the Theory of Numbers

INTRODUCTION

A simple theory of invariants for the modular forms and linear transformations employed in the theory of numbers should be of an importance commensurate with that of the theory of invariants in modern algebra and analytic projective geometry, and should have the advantage of introducing into the theory of numbers methods uniform with those of algebra and geometry.

In considering the invariants of a modular form (a homogeneous polynomial with integral coefficients taken modulo p, where p is a prime), we see at once that the rational integral invariants of the corresponding algebraic form with arbitrary variables as coefficients give rise to as many modular invariants of the modular form, and that there are numerous additional invariants peculiar to the case of the theory of numbers. Moreover, nearly all of the processes of the theory of algebraic invariants, whether symbolic or not, either fail for modular invariants or else become so complicated as to be useless. For instance, the annihilators are no longer linear differential operators. The attempt to construct a simple theory of modular invariants from the standpoints in vogue in the algebraic theory was a failure, although useful special results were obtained in this laborious way. Later I discovered a new standpoint which led to a remarkably simple theory of modular invariants. This standpoint is of function-theoretic character, employing the values of the invariant, and using linear transformations only in the preliminary problem of separating into classes the particular forms obtained by assigning special values to the coefficients of the ground form. As to the practical value of the new theory as a working tool, it may be observed that the problem to find a fundamental system of modular seminvariants of a binary form is from the new standpoint a much simpler problem than the corresponding one in the algebraic case; indeed, we shall exhibit explicitly the fundamental system of modular seminvariants for a binary form of general degree.

It will now be clear why these Lectures make no use of the technical theories of algebraic invariants. On the contrary, they afford an introduction to that subject from a new standpoint and, in particular, throw considerable new light on the relations between the subjects of rational integral invariants and transcendental invariants of algebraic forms and the corresponding questions for seminvariants. Again, I shall make no use of technical theory of numbers, presupposing merely the concepts of congruence and primitive roots, Fermat's theorem, and (in Lectures III and V) the concept of quadratic residues.

The developments given in these Lectures are new, with exceptions in the case of Lecture I, which presents an introduction to the theory, and in the case of the earlier and final sections of Lecture III. But in these cases the exposition is considerably simpler and more elementary than that in my published papers on the same topics. The contacts with the work of other writers will be indicated at the appropriate places. Much light is thrown upon the unsolved problem of Hurwitz concerning formal invariants.

In many parts of these Lectures, I have not aimed at complete generality and exhaustiveness, but rather at an illumination of typical and suggestive topics, treated by that particular method which I have found to be the best of various possible methods. Surely in a new subject in which most of the possible methods are very complex, it is desirable to put on record an account of the simple successful methods. Finally, it may be remarked that the present theory is equally simple when the coefficients of the forms and linear transformations are not integers, but are elements of any finite field.

I am much indebted to Dr. Sanderson and Professors Cole and Glenn for reading the proof sheets.

LECTURE I

A THEORY OF INVARIANTS APPLICABLE TO ALGEBRAIC AND MODULAR FORMS

INTRODUCTION TO THE ALGEBRAIC SIDE OF THE THEORY BY MEANS OF THE EXAMPLE OF AN ALGEBRAIC QUADRATIC FORM IN m VARIABLES, §§ 1–3

1. *Classes of Algebraic Quadratic Forms.*—Let the coefficients of

$$q_m = \sum_{i,j=1}^{m} \beta_{ij} x_i x_j \qquad (\beta_{ji} = \beta_{ij}) \tag{1}$$

be ordinary real or complex numbers. Let the determinant

$$D = |\beta_{ij}| \qquad (i, j = 1, \cdots, m) \tag{2}$$

of a particular form q_m be of rank r $(r > 0)$; then every minor of order exceeding r is zero, while at least one minor of order r is not zero. There exists a linear transformation of determinant unity which replaces this q_m by a form*

$$\alpha_1 x_1^2 + \cdots + \alpha_r x_r^2 \quad (\alpha_1 \neq 0, \cdots, \alpha_r \neq 0). \tag{3}$$

Indeed, if $\beta_{11} \neq 0$, we obtain a form lacking $x_1 x_2, \cdots, x_1 x_m$ by substituting

$$x_1 - \beta_{11}^{-1}(\beta_{12} x_2 + \cdots + \beta_{1m} x_m)$$

for x_1. If $\beta_{11} = 0$, $\beta_{ii} \neq 0$, we substitute x_i for x_1 and $-x_1$ for x_i; while, if every $\beta_{kk} = 0$, and $\beta_{12} \neq 0$, we substitute $x_2 + x_1$ for x_2; in either case we obtain a form in which the coefficient of x_1^2 is not zero. We now have $\alpha_1 x_1^2 + \phi$, where $\alpha_1 \neq 0$ and ϕ involves only $x_2, \cdots, x_m$. Proceeding similarly with ϕ, we ultimately obtain a form (3).

Now (3) is replaced by a similar form having $\alpha_1 = 1$ by the

* Note for later use that each α_k and each coefficient of the transformation is a rational function of the β's with integral coefficients.

transformation

$$x_1 = \alpha_1^{-\frac{1}{2}} x_1', \quad x_m = \alpha_1^{\frac{1}{2}} x_m', \quad x_i = x_i' \quad (i = 2, \cdots, m-1)$$

of determinant unity. Hence there exists a linear transformation with complex coefficients of determinant unity which replaces q_m by

$$(4) \qquad x_1^2 + \cdots + x_{m-1}^2 + D x_m^2, \qquad x_1^2 + \cdots + x_r^2,$$

according as $r = m$ or $r < m$. In the first case, the final coefficient is D since the determinant (2) of a form q_m equals that of the form derived from q_m by any linear transformation of determinant unity. Hence all quadratic forms (1) may be separated into the classes

$$(5) \qquad C_{m, D}, \quad C_r \quad (D \neq 0, r = 0, 1, \cdots, m-1),$$

where, for a particular number $D \neq 0$, the class $C_{m, D}$ is composed of all forms q_m of determinant D, each being transformable into (4_1); while, for $0 < r < m$, the class C_r is composed of all forms of rank r, each being transformable into (4_2); and, finally, the class C_0 is composed of the single form with every coefficient zero. In the last case, the determinant D is said to be of rank zero. Using also the fact that the rank of the determinant of a quadratic form is not altered by linear transformation, we conclude that *two quadratic forms are transformable into each other by linear transformations of determinant unity if and only if they belong to the same class* (5).

2. *Single-valued Invariants of q_m.*—Using the term function in Dirichlet's sense of correspondence, we shall say that a single-valued function ϕ of the undetermined coefficients β_{ij} of the general quadratic form q_m is an *invariant* of q_m if ϕ has the same value for all sets β_{ij}', β_{ij}'', $\cdots$ of coefficients of forms q_m', q_m'', $\cdots$ belonging to the same class.* The values $v_{m, D}$, v_r of ϕ for the various classes (5) are in general different. For example, the determinant D is an invariant; likewise the single-valued func-

* Briefly, if ϕ has the same value for all forms in any class.

tion r of the undetermined coefficients β_{ij} which specifies the rank of $|\beta_{ij}|$.

Each consistent set of values of D and r uniquely determines a class (5) and, by definition, each class uniquely determines a value of ϕ. Hence ϕ is a single valued function of D and r.

Every single-valued invariant of a system of forms is a single-valued function of the invariants (D *and* r *in our example*) *which completely characterize the classes.*

3. *Rational Integral Invariants of* q_m.—If the invariant ϕ is a rational integral function of the coefficients β_{ij}, it equals a rational integral function of D. For, if the β's have any values such that $D \neq 0$, ϕ has the same value for the form (1) as for the particular form (4_1) of the same class. Hence $\phi = P(D)$, where $P(D)$ is a polynomial in D with numerical coefficients. Since this equation holds for all sets of β's whose determinant is not zero, it is an identity.

Introduction to the Number Theory Side of the Theory of Invariants by Means of the Example of a Modular Quadratic Form, §§ 4–7

4. *Classes of Modular Quadratic Forms* q_m.—Let $x_1, \cdots, x_m$ be indeterminates in the sense of Kronecker. Let each β_{ij} be an integer taken modulo p, where p is an odd prime. Then the expression (1) is called a modular quadratic form. By § 1, there exists a linear transformation, whose coefficients are integers* taken modulo p and whose determinant is congruent to unity, which replaces q_m by a quadratic form (3) in which each α_k is an integer not divisible by p. Thus† each α_k is congruent to a power of a primitive root ρ of p. After applying a linear transformation of determinant unity which permutes $x_1^2, \cdots, x_r^2$, we may assume that $\alpha_1, \cdots, \alpha_s$ are even powers of ρ and that $\alpha_{s+1}, \cdots, \alpha_r$ are odd powers of ρ. The transformation which

* Perhaps initially of the form a/b, where a and b are integers, b not divisible by p. But there exists an integral solution q of $qb \equiv a \pmod{p}$.

† For $p = 5$, $\rho = 2$, $1 \equiv 2^4$, $2 \equiv 2^1$, $3 \equiv 2^3$, $4 \equiv 2^2 \pmod{5}$.

multiplies a particular x_i $(i < m)$ by ρ^k and x_m by ρ^{-k} is of determinant unity.

First, let $r < m$. Applying transformations of the last type to (3), we obtain

$$(6) \qquad x_1^2 + \cdots + x_s^2 + \rho x_{s+1}^2 + \cdots + \rho x_r^2.$$

Under the transformation of determinant unity

$$x_i = \alpha X_i + \beta X_j, \quad x_j = -\beta X_i + \alpha X_j, \quad x_m = (\alpha^2 + \beta^2)^{-1} X_m,$$

$x_i^2 + x_j^2$ becomes $(\alpha^2 + \beta^2)(X_i^2 + X_j^2)$. Choose* integers α, β so that

$$(7) \qquad \rho(\alpha^2 + \beta^2) \equiv 1 \qquad (\text{mod } p).$$

Hence the sum of two terms of (6) with the coefficient ρ can be transformed into a sum of two squares. Thus by means of a linear transformation, with integral coefficients of determinant unity, q_m can be reduced to one of the forms

$$(8) \quad x_1^2 + \cdots + x_{r-1}^2 + x_r^2, \quad x_1^2 + \cdots + x_{r-1}^2 + \rho x_r^2 \quad (0 < r < m).$$

Next, let $r = m$. We obtain initially

$$x_1^2 + \cdots + x_s^2 + \rho x_{s+1}^2 + \cdots + \rho x_{m-1}^2 + \sigma x_m^2,$$

in which σ need not equal ρ as in (6). If there be an even number of terms with the coefficient ρ, we obtain as above a form of type (4_1). In the contrary case, we get

$$f = x_1^2 + \cdots + x_{m-2}^2 + \rho x_{m-1}^2 + \rho^{-1} D x_m^2.$$

If $D \equiv \rho^{2l+1}$ (mod p), f is transformed into (4_1) by

$$x_{m-1} = -\rho^l X_m, \quad x_m = \rho^{-l} X_{m-1}.$$

But if $D \equiv \rho^{2l}$, f is reduced to (4_1) by the transformation

$$x_{m-1} = \alpha X_{m-1} + \delta \rho^{2l-1} X_m, \quad x_m = -\delta X_{m-1} + \alpha \rho X_m,$$

$$\rho(\alpha^2 + \rho^{2l-2}\delta^2) \equiv 1,$$

---

* If $p = 5$, $\rho = 2$, we may take $\alpha = \beta = 2$. For any p, either there is an integer l such that $l^2 \equiv -1$ (mod p) and we may take $\rho(\alpha + l\beta) \equiv 1$, $\alpha - l\beta \equiv 1$; or else $x^2 + 1$ takes $1 + (p - 1)/2$ incongruent values modulo p, no one divisible by p, when x ranges over the integers $0, 1, \cdots, p - 1$, so that $x^2 + 1$ takes at least one value of the form ρ^{2e-1}. In the latter event, $\alpha = \rho^{-e}$, $\beta = x\alpha$ satisfy (7).

of determinant unity. The final condition is of the form (7) with $\beta = \rho^{l-1}\delta$ and hence has integral solutions α, δ.

Hence the classes of modular quadratic forms are

$$C_{m,\,D},\quad C_{r,\,1},\quad C_{r,\,-1},\quad C_0 \qquad (D = 1, \cdots, p-1;\ r = 1, \cdots, m-1), \tag{9}$$

where $C_{m,\,D}$ is composed of all modular quadratic forms whose determinant is a given integer D not divisible by p, each being transformable into (4_1), where $C_{r,\,1}$ and $C_{r,\,-1}$ are composed of all forms transformable into (8_1) and (8_2) respectively, and C_0 is composed of the form all of whose coefficients are zero.

Two modular quadratic forms are transformable into each other by linear transformations with integral coefficients of determinant unity modulo p if and only if they belong to the same class (9). Indeed, since D and r are invariants,* it remains only to show that the two forms (8) are not transformable into each other.† But if a linear transformation

$$x_i = \sum_{j=1}^{m} \alpha_{ij} X_j \qquad (i = 1, \cdots, m)$$

replaces $f = x_1^2 + \cdots + x_r^2$ by $F = X_1^2 + \cdots + X_{r-1}^2 + \rho X_r^2$, then, for $j > r$,

$$\frac{\partial f}{\partial X_j} \equiv 2 \sum_{i=1}^{r} x_i \frac{\partial x_i}{\partial X_j} = \frac{\partial F}{\partial X_j} = 0, \quad \frac{\partial x_i}{\partial X_j} = \alpha_{ij} = 0 \qquad (i \leqq r, j > r),$$

$$x_i = \sum_{j=1}^{r} \alpha_{ij} X_j \qquad (i = 1, \cdots, r).$$

Hence under this partial transformation on $x_1, \cdots, x_r$, we would have $f = F$. Thus the determinant of F would equal $|\alpha_{ij}|^2$ times the determinant unity of f and hence equal an even power of ρ. But the determinant of F is actually ρ.

* r is now the maximum order of a minor not divisible by p.

† An immediate proof follows from the values taken by the invariant A_r given below. But as the necessity of constructing A_r is based upon the fact that the forms (8) do not belong to the same class, it seems preferable to prove the last fact without the use of A_r.

The invariants D and r therefore do not completely characterize the classes of modular quadratic forms, a result in contrast to that for algebraic quadratic forms. We shall give a criterion to decide whether a given form of rank r $(0 < r < m)$ is of class $C_{r,1}$ or of class $C_{r,-1}$ and later deduce an invariantive criterion.

5. *Criterion for Classes* $C_{r,\pm 1}$.—Such a criterion may be obtained from Kronecker's elegant theory of quadratic forms.* We shall make use of the theorem that a symmetrical determinant of rank r $(r > 0)$ has a non-vanishing principal minor M of order r, i. e., one whose diagonal elements lie in the main diagonal of the given determinant.† After an evident linear transformation of determinant unity, we may set

$$(10)\qquad M = |\beta_{ij}| \not\equiv 0 \pmod{p} \qquad (i, j = 1, \cdots, r).$$

In the present problem, $r < m$. To q_m apply the transformation

$$x_i = X_i + c_iX_m \qquad (i = 1, \cdots, r),$$
$$x_i = X_i \qquad (i = r + 1, \cdots, m)$$

of determinant unity in which the c_i are integers. We get

$$\sum_{i,j=1}^{m-1} \beta_{ij}X_iX_j + 2\sum_{j=1}^{m-1} B_{jm}X_jX_m + \left(\sum_{j=1}^{r} B_{jm}c_j + B_{mm}\right)X_m^2,$$

where

$$B_{jm} = \sum_{i=1}^{r} \beta_{ij}c_i + \beta_{jm} \qquad (j = 1, \cdots, m).$$

In view of (10) there are unique values of $c_1, \cdots, c_r$ such that

$$B_{jm} \equiv 0 \pmod{p} \qquad (j = 1, \cdots, r).$$

But the determinant of the coefficients of $c_1, \cdots, c_r, 1$ in

$$B_{1m},\ B_{2m},\ \cdots,\ B_{rm},\ B_{km} \qquad (r < k \leqq m)$$

* Kronecker, Werke, vol. 1, p. 166, p. 357; cf. Gundelfinger, *Crelle*, vol. 91 (1881), p. 221; Bôcher, Introduction to Higher Algebra, p. 58, p. 139.

† The most elementary proof is that by Dickson, *Annals of Mathematics*, ser. 2, vol. 15 (1913), pp. 27, 28. For other short proofs, see Wedderburn, ibid., p. 29, and Kowalewski, Determinantentheorie, pp. 122–124.

is the minor of β_{km} in the determinant

$$|\beta_{ij}| \qquad (i, j = 1, \cdots, r, k, m)$$

and hence is zero, being of order $r + 1$. Hence $B_{km} \equiv 0$. Thus q_m has been transformed into

$$\sum_{i,j=1}^{m-1} \beta_{ij} X_i X_j.$$

After repetitions of this process, q_m is transformed into*

$$\sum_{i,j=1}^{r} \beta_{ij} x_i x_j. \tag{11}$$

This form, of determinant M, can be reduced (§ 4) to

$$x_1^2 + \cdots + x_{r-1}^2 + M x_r^2$$

by a linear transformation on $x_1, \cdots, x_r$ with integral coefficients of determinant unity modulo p. Express M as a power $\rho^{2l+\epsilon}$ ($\epsilon = 0$ or 1) of a primitive root. Since $r < m$, we may replace x_r by $\rho^{-l} x_r$ and x_m by $\rho^l x_m$ and obtain (8_1) or (8_2) according as $\epsilon = 0$ or $\epsilon = 1$. Now $\rho^{(p-1)/2}$ is not congruent to unity, but its square is congruent to unity modulo p, by Fermat's theorem; hence it is $\equiv -1$. Thus, in the respective cases,

$$M^{\frac{p-1}{2}} \equiv +1 \quad \text{or} \quad -1 \qquad (\text{mod } p). \tag{12}$$

Hence if a form is of rank r and if M is any chosen r-rowed principal minor not divisible by p, the form is of class $C_{r,1}$ or $C_{r,-1}$ according as the first or second alternative (12) holds.

6. *Invariantive Criterion for Classes* $C_{r,\pm1}$.—A function which has the value $+1$ for any form of class $C_{r,+1}$, the value -1 for any form of class $C_{r,-1}$, and the value zero for the remaining classes $C_{m,D}$, C_0, $C_{k,\pm1}$ ($k \neq r$), is an invariant (§ 2). This function† is

* This proof and the results in §§ 4–13 are due to Dickson, *Transactions of the American Mathematical Society*, vol. 10 (1909), pp. 123–133.

† Constructed synthetically in the paper last cited.

$$(13)\quad \begin{aligned} A_r = \{&M_1^{\frac{p-1}{2}} + M_2^{\frac{p-1}{2}}(1 - M_1^{p-1}) + \cdots \\ &+ M_n^{\frac{p-1}{2}}(1 - M_1^{p-1}) \cdots (1 - M_{n-1}^{p-1})\}\Pi(1 - d^{p-1}), \end{aligned}$$

where $M_1, \cdots, M_n$ denote the principal minors of order r taken in any sequence, and d ranges over the principal minors of orders exceeding r. For, if any $d \not\equiv 0$, the rank exceeds r and $A_r \equiv 0$ by Fermat's theorem. Next, let every $d \equiv 0$, so that the rank is r or less, and the final factor in (13) is congruent to unity. Then, if every $M_i \equiv 0$, the rank is less than r and $A_r \equiv 0$. But, if $M_1 \not\equiv 0$,

$$A_r \equiv M_1^{\frac{p-1}{2}} \equiv \pm 1 \qquad (\text{mod } p),$$

by (12), the sign being the same as in $C_{r,\,\pm 1}$. If $M_1 \equiv 0$, $M_2 \not\equiv 0$,

$$A_r \equiv M_2^{\frac{p-1}{2}} \equiv \pm 1 \qquad (\text{mod } p),$$

etc. Note for later use that

$$(14)\qquad A_m = D^{\frac{p-1}{2}}.$$

7. *Rational Integral Invariants of q_m.*—The function

$$(15)\qquad I_0 = \Pi(1 - \beta_{ij}^{p-1}) \qquad (i, j = 1, \cdots, m;\ i \leqq j)$$

has the value 1 for the form (of class C_0) all of whose coefficients are zero and the value 0 for all remaining forms q_m, and hence is an invariant of q_m. We now have rational integral invariants

$$(16)\qquad D, \quad A_1, \quad \cdots, \quad A_{m-1}, \quad I_0$$

which completely characterize the classes (9). Hence, by the general theorem in § 12, any rational integral invariant of the modular form q_m is a rational integral function of the invariants (16) with integral coefficients. In other words, invariants (16) form a fundamental system of rational integral invariants of q_m.

If we employ not merely, as before, linear transformations with integral coefficients of determinant unity modulo p, but those of all determinants, we obtain at once the classes

$$C_{r,\,\pm 1}, \quad C_0 \qquad\qquad (r = 1, \cdots, m),$$

and see that these are characterized by $A_1, \cdots, A_m, I_0$. The latter therefore form a fundamental system of rational integral absolute invariants. But D is a relative invariant.

General Theory of Modular Invariants, §§ 8–14

8. *Definitions.*—Let S be any system of forms in $x_1, \cdots, x_m$ with undetermined integral coefficients taken modulo p, a prime. Let G be any group of linear transformations on $x_1, \cdots, x_m$ with integral coefficients taken modulo p. The particular systems $S', S'', \cdots$, obtained from S by assigning to the coefficients particular sets of integral values modulo p, may be separated into *classes* $C_0, C_1, \cdots, C_{n-1}$ such that two systems belong to the same class if and only if they are transformable into each other by transformations of G.

A single-valued function ϕ of the coefficients of the forms in the system S is called an *invariant* of S under G if, for $i = 0, 1, \cdots, n-1$, the function ϕ has the same value v_i for all systems of forms in the class C_i.

In case the values taken by ϕ are integers which may be reduced at will modulo p and congruent values be identified, the invariant is called *modular*. Since this reduction can be effected on each coefficient of the modular forms comprising our system S, any rational integral invariant of S is a modular invariant.

An example of a non-modular invariant is the transcendental function r defining the rank of the determinant of the modular quadratic form q_m. The values of r are evidently not to be identified when merely congruent modulo p. However, the residue of r modulo p is a modular invariant, since

$$(17) \qquad r \equiv A_1^2 + 2A_2^2 + \cdots + mA_m^2 \pmod{p}.$$

9. *Modular Invariants are Rational and Integral.*—Any modular invariant ϕ of a system S of modular forms can be identified with a rational integral function (with integral coefficients) of the coefficients $c_1, \cdots, c_s$ appearing in the forms of the system S.

For, if

$$\phi \equiv v_{e_1, \cdots, e_s} \quad \text{when} \quad c_1 \equiv e_1, \cdots, c_s \equiv e_s \qquad (\text{mod } p),$$

then ϕ is identically congruent (as to $c_1, \cdots, c_s$) to

$$\sum_{e_1, \cdots, e_s = 0}^{p-1} v_{e_1, \cdots, e_s} \prod_{i=1}^{s} \{1 - (c_i - e_i)^{p-1}\}, \tag{18}$$

as shown by Fermat's theorem.

10. *Characteristic Modular Invariants.*—The characteristic invariant I_k of the class C_k is defined to be that modular invariant which has the value unity for systems of forms of the class C_k and the value zero for any of the remaining classes.

For example, for a single quadratic form q_m, I_0 is given by (15), while the characteristic invariants for the classes $C_{r,\,1}$ and $C_{r,\,-1}$ are

$$I_{r,\,1} = \tfrac{1}{2}(A_r^2 + A_r), \quad I_{r,\,-1} = \tfrac{1}{2}(A_r^2 - A_r). \tag{19}$$

For any system of forms with the coefficients $c_1, \cdots, c_s$, we have

$$I_k = \sum \prod_{i=1}^{s} \{1 - (c_i - c_i^{(k)})^{p-1}\}, \tag{20}$$

where the sum extends over all sets of coefficients $c_1^{(k)}, \cdots, c_s^{(k)}$ of the various systems of forms of class C_k. In particular, in accord with (15),

$$I_0 = \prod_{i=1}^{s} (1 - c_i^{p-1}). \tag{21}$$

11. *Number of Linearly Independent Modular Invariants.*—Since any modular invariant I takes certain values $v_0, \cdots, v_{n-1}$ for the respective classes $C_0, \cdots, C_{n-1}$, we have

$$I = v_0 I_0 + v_1 I_1 + \cdots + v_{n-1} I_{n-1}. \tag{22}$$

Hence any modular invariant can be expressed in one and but one way as a linear homogeneous function of the characteristic invariants. Moreover, *the number of linearly independent modular invariants equals the number of classes.*

For example, using (19), we see that a complete set of linearly independent modular invariants of the quadratic form q_m modulo p $(p > 2)$ is given by

$$(23)\quad I_0,\ A_r,\ A_r^2\ (r = 1, \cdots, m-1),\quad D^k\ (k = 1, \cdots, p-1).$$

12. *Fundamental Systems of Modular Invariants.*—While, by (22), the characteristic invariants $I_0, \cdots, I_{n-1}$ form a fundamental system of modular invariants of a system S of modular forms, it is usually much easier to find another fundamental system. In fact, certain invariants are usually known in advance, e. g., the invariants of the corresponding system of algebraic forms. We shall prove the following fundamental theorem:

If the modular invariants A, B, $\cdots$, L completely characterize the classes, they form a fundamental system of modular invariants.

For example, $I_0, \cdots, I_{n-1}$ evidently completely characterize the classes and were seen to form a fundamental system.

Let $c_1, \cdots, c_s$ be the coefficients of the forms in the system S. Let each c_i take the values $0, 1, \cdots, p-1$. For the resulting p^s sets of values of the c's, let the rational integral functions $A, B, \cdots, L$ of $c_1, \cdots, c_s$ take the distinct sets of values

$$A_i,\quad B_i,\quad \cdots,\quad L_i\quad (i = 0, \cdots, n-1).$$

Thus there are n classes of systems S and by hypothesis the ith class is uniquely defined by the values $A_i, \cdots, L_i$ of our invariants. A rational integral invariant $\phi(c_1, \cdots, c_s)$ takes the same value for all systems of forms in the ith class, so that this value may be designated by ϕ_i. Now the polynomial

$$P(A, B, \cdots, L) = \sum_{i=0}^{n-1} \phi_i\{1 - (A - A_i)^{p-1}\} \cdots \{1 - (L - L_i)^{p-1}\}$$

is congruent to ϕ_i when $A \equiv A_i, \cdots, L \equiv L_i \pmod{p}$. Hence

$$\phi(c_1, \cdots, c_s) \equiv P(A, B, \cdots, L) \pmod{p}$$

for all sets of integral values of $c_1, \cdots, c_s$. In view of Fermat's theorem, we may assume that each exponent in $\phi(c_1, \cdots, c_s)$ is less than p. If we replace $A, \cdots, L$ by their expressions in terms of the c's, $P(A, \cdots, L)$ becomes a polynomial, which, after exponents are reduced below p, will be designated by $\psi(c_1, \cdots, c_s)$. Then ϕ and ψ are identically congruent in $c_1, \cdots, c_s$, that is, corresponding coefficients are congruent modulo p. In fact, a polynomial of type ϕ is uniquely determined by its values for the p^s sets of values of $c_1, \cdots, c_s$, each chosen from $0, 1, \cdots, p-1$ (§ 9). Hence ϕ can be expressed as a polynomial in $A, \cdots, L$ with integral coefficients.*

13. *Minor Rôle of Modular Covariants.*—In contrast with the case of algebraic forms, the classes of modular forms are completely characterized by rational integral invariants. Such invariants therefore suffice to express all invariantive properties of a system of modular forms. In this respect, modular covariants play a superfluous rôle. For instance, a projective property of a system of algebraic forms is often expressed by the identical vanishing of a covariant. But if C is a modular covariant with the coefficients $c_1, \cdots, c_s$, then I_0 given by (21) is a modular invariant of C and hence of the initial system of forms. We have $C \equiv 0$ or $C \not\equiv 0 \pmod{p}$ identically, according as $I_0 \equiv 1$ or $I_0 \equiv 0$.

14. *References to Further Developments.*—This general theory of modular invariants has been applied by me to determine a complete set of linearly independent modular invariants of q linear forms on m variables,† and a fundamental system of modular invariants of a pair of binary quadratic forms and of a pair of binary forms, one quadratic and the other linear.‡

The theory has been extended to combinants and applied to a pair of binary quadratic forms.§

* This correct theorem for any finite field cannot be extended at once to any field as stated by me in *American Journal of Mathematics*, vol. 31 (1909), top of p. 338.

† *Proceedings of the London Mathematical Society*, ser. 2, vol. 7 (1909), pp. 430–444.

‡ *American Journal of Mathematics*, vol. 31 (1909), pp. 343–354; cf. pp. 103–146, where a less effective method is used.

§ Dickson, *Quarterly Journal of Mathematics*, vol. 40 (1909), pp. 349–366.

LECTURE II

SEMINVARIANTS OF ALGEBRAIC AND MODULAR BINARY FORMS

Introductory Example of the Binary Quartic Form, §§ 1–6

1. *Comparative View.*—Let the forms

$$f = a_0x^4 + 4a_1x^3y + 6a_2x^2y^2 + 4a_3xy^3 + a_4y^4,$$

with real or complex coefficients, be separated into classes such that two forms f are transformable into one another by a transformation of type

$$(1) \qquad x = x' + ty', \quad y = y',$$

if and only if they belong to the same class. Then a single-valued function $S(a_0, \cdots, a_4)$ is called a seminvariant of f if it has the same value for all of the forms in any class.

By the repeated application of this definition and without the aid of new principles, we shall obtain a fundamental system of rational integral seminvariants of f, then on the one hand the additional single-valued seminvariant needed to form with these a fundamental system of single-valued seminvarints, and on the other hand the additional rational integral modular seminvariants needed to form with them a fundamental system of modular seminvariants of f. It is such a comparative view that we desire to emphasize here. In later sections, we shall show that it is usually much simpler to treat the modular case independently and in particular without introducing all of the algebraic seminvariants, which become very numerous and most unwieldy for forms of high degree. The rational integral seminvariants S of an algebraic form are of special importance since each is the leading coefficient of one and but one covariant, which can be found from S by a process of differentiation. For example, the seminvariant a_0 is the leading coefficient of the covariant f.

2. *The Classes of Algebraic Quartic Forms.*—Consider a quartic form f in which a_k is the first non-vanishing coefficient. Apply transformation (1) with

$$t = \frac{-a_{k+1}}{(k+1)a_k}. \tag{2}$$

We obtain a form having zero in place of the former a_{k+1}. Dropping the accents on x', y', we obtain, for $k = 0, 1, 2, 3$, the respective forms

$$a_0 \neq 0: \qquad a_0x^4 + 6a_0^{-1}S_2x^2y^2 + 4a_0^{-2}S_3xy^3 + a_0^{-3}S_4y^4, \tag{3}$$

$$a_0 = 0, \quad a_1 \neq 0: \qquad 4a_1x^3y + a_1^{-1}S_{13}xy^3 + a_1^{-2}S_{14}y^4, \tag{4}$$

$$a_0 = a_1 = 0, \quad a_2 \neq 0: \qquad 6a_2x^2y^2 + \tfrac{1}{3}a_2^{-1}S_{24}y^4, \tag{5}$$

$$a_0 = a_1 = a_2 = 0, \quad a_3 \neq 0: \qquad 4a_3xy^3, \tag{6}$$

$$a_0 = a_1 = a_2 = a_3 = 0: \qquad a_4y^4, \tag{7}$$

no transformation having been made in the last case. Here

$$S_2 = a_0a_2 - a_1^2, \quad S_3 = a_0^2a_3 - 3a_0a_1a_2 + 2a_1^3, \tag{8}$$

$$S_4 = a_0^3a_4 - 4a_0^2a_1a_3 + 6a_0a_1^2a_2 - 3a_1^4, \tag{9}$$

$$\begin{gathered} S_{13} = 4a_1a_3 - 3a_2^2, \quad S_{14} = a_1^2a_4 - 2a_1a_2a_3 + a_2^3, \\ S_{24} = 3a_2a_4 - 2a_3^2. \end{gathered} \tag{10}$$

If we apply to one of the forms (3)–(6) a transformation (1) with $t \neq 0$, we obtain a form having an additional (second) term. Hence no two of the forms (3)–(7) can be transformed into each other by a transformation (1), so that each represents a class of forms. For example, there is a class (5) for each set of values of the parameters a_2 and S_{24} ($a_2 \neq 0$).

3. *Rational Integral Seminvariants of an Algebraic Quartic.*—First, a_0 is a seminvariant since it has a definite value $\neq 0$ for any form in any class (3) and the value zero for any form in any class (4)–(7). Next, S_2, S_3, S_4 are seminvariants, since they have constant values

$$S_2 = -a_1^2, \quad S_3 = 2a_1^3, \quad S_4 = -3a_1^4 \qquad (\text{if } a_0 = 0) \tag{11}$$

for any form in any class (4)–(7), and constant values for any form of a definite class (3), for which therefore a_0 has a definite value $\neq 0$ and $a_0^{-1}S_2, \cdots$, and hence each S_i, has a definite value. Moreover, *these seminvariants* a_0, S_2, S_3, S_4 *completely characterize the classes* (3).

Consider a quartic form f in which a_0, a_1, a_2, a_3, a_4 are arbitrary, except that $a_0 \neq 0$. Any rational integral seminvariant $S(a_0, \cdots, a_4)$ has the same value for f as for the particular form (3) in the same class as f. Hence

$$S = S\left(a_0, 0, \frac{S_2}{a_0}, \frac{S_3}{a_0^2}, \frac{S_4}{a_0^3}\right) = \frac{\phi(a_0, S_2, S_3, S_4)}{a_0^j},$$

where ϕ is a rational integral function of its arguments. We therefore seek such functions ϕ as are divisible by a power of a_0, and hence by (11) in which the terms involving only a_1 cancel. The function of lowest degree is evidently

$$(12) \qquad S_4 + 3S_2^2 = a_0^2 I, \quad I \equiv a_0a_4 - 4a_1a_3 + 3a_2^2.$$

The next lowest degree is 6 and the function is

$$dS_2S_4 + eS_3^2 + (3d + 4e)S_2^3.$$

The coefficient of d is $a_0^2IS_2$, that of e is

$$(13) \qquad \begin{gathered} S_3^2 + 4S_2^3 = a_0^2D \\ (D \equiv a_0^2a_3^2 - 6a_0a_1a_2a_3 + 4a_0a_2^3 + 4a_1^3a_3 - 3a_1^2a_2^2). \end{gathered}$$

Hence for $d = 1, e = -1$, the function is the product of a_0^2 and

$$(14) \quad IS_2 - D = a_0J, \quad J \equiv a_0a_2a_4 - a_0a_3^2 + 2a_1a_2a_3 - a_1^2a_4 - a_2^3.$$

We do not retain D since it is expressible in terms of the other functions. Eliminating D between (13) and (14), we get

$$(15) \qquad S_3^2 + 4S_2^3 - a_0^2IS_2 + a_0^3J \equiv 0.$$

Now I and J are seminvariants. Indeed, if $a_0 \neq 0$, they are expressible in terms of the parameters a_0, S_i in (3) and hence each has the same value for any form in a class (3); while

$$(16) \qquad I = -S_{13}, \quad J = -S_{14} \qquad \text{(if } a_0 = 0\text{)},$$

so that each has the same value for any form in a class (4); finally,

$$I = 3a_2^2, \quad J = -a_2^3 \qquad (\text{if } a_0 = a_1 = 0), \tag{17}$$

so that each has the same value for any form in a class (5)–(7).

From ϕ we eliminate S_4 by means of (12) and then the second and higher powers of S_3 by means of (15). Thus S equals N/a_0^k, where N is a rational integral function of

$$a_0, \quad S_2, \quad S_3, \quad I, \quad J, \tag{18}$$

of degree 0 or 1 in S_3. If $k > 0$, we may evidently assume that not every term of the polynomial N in the arguments (18) has the factor a_0. Let $P(S_2, S_3, I, J)$ denote the aggregate of the terms of N not involving a_0 explicitly. We shall prove that, if $k > 0$, N/a_0^k is then not a rational integral function of $a_0, \cdots, a_4$. For, if it be, P vanishes when $a_0 = 0$. By (11) and (16), the terms independent of a_0 in J involve a_4, while those in I, S_2, S_3 do not. Hence J does not occur in P. Then, by (11) and the term $3a_2^2$ in I, we conclude that I does not occur in P. Thus P is a polynomial in S_2 and S_3 of degree 0 or 1 in S_3 and is not identically zero. By (11), it cannot vanish for $a_0 = 0$.

Under the initial assumption that $a_0 \neq 0$, we have now proved that any rational integral seminvariant S equals a polynomial in the functions (18). The resulting equality is therefore an identity.

The seminvariants (18) *form a fundamental system of rational integral seminvariants of the algebraic quartic form.**

They are connected by the relation, or syzygy, (15).

4. *Invariantive Characterization of the Classes.*—By § 3, the classes (3) are completely characterized by the seminvariants a_0, S_2, S_3, I. These with J characterize the classes (4) having $a_0 = 0$, $a_1 \neq 0$. For, by (11), S_2 and S_3 determine a_1; while, by (16), I and J determine the remaining parameters in (4).

* The above proof differs from that by Cayley in minor details and in the method of obtaining the functions (18) and the verification that they are seminvariants (the present method being based upon the classes).

The parameter a_2 $(a_2 \neq 0)$ in (5) is determined by I and J, in view of (17).

We have now gone as far as is possible in the characterization of the classes by means of rational integral seminvariants S, since the parameters S_{24}, a_3, a_4 in (5)–(7) cannot be determined by such seminvariants. Indeed,* for $a_0 = a_1 = 0$, we have $S_2 = S_3 = 0$ by (11), while I and J reduce to powers of a_2 by (17).

5. *Single-valued Seminvariants.*—We may, however, construct a single-valued seminvariant which shall determine these outstanding parameters S_{24}, a_3, a_4. To this end consider the single-valued function V defined as follows by its values in the sense of Dirichlet. We take $V = 0$ if $a_0 \neq 0$ or if $a_1 \neq 0$, and $V = S_{24}$, a_3, a_4 in the respective cases (5), (6), (7). Since V has the same value for all forms in any class, it is a seminvariant. The seminvariants (18) and V completely characterize the classes (3)–(7) and hence, by § 2 of Lecture I, form a fundamental system of single-valued seminvariants of the algebraic binary quartic form.

6. *Seminvariants of a Modular Quartic Form.*—Passing to the number theory case, let the coefficients of the quartic form f be integers taken modulo p, where p is a prime exceeding 3. The denominator in (2) is then not divisible by p, so that the classes are again (3)–(7).

By the general theory in Lecture I, it is possible to characterize all of the classes by means of rational integral seminvariants, and the latter will then form a fundamental system. In particular, we do not now require the use of such a bizarre function as that used in § 5.

* A proof of this fact, not based upon the final theorem of § 3, would afford a better insight into the nature of the last steps in § 3 and explain, in particular, why we stopped with I and J and did not consider combinations of the S_i of higher than the sixth degree in the a's. To this end, let S be a seminvariant homogeneous of total degree i, in the a's, and isobaric, of constant weight w. As well known, $4i \geqq 2w$. Thus S cannot have a term a_3^l or a_4^l and cannot reduce, when $a_0 = a_1 = 0$, to $a_2^l S_{24}^m$ $(m > 0)$, of degree $l + 2m$ and weight $2l + 6m$.

We shall make frequent use of the abbreviation

(19) $$P_i = (1 - a_0^{p-1})(1 - a_1^{p-1}) \cdots (1 - a_i^{p-1}).$$

Then P_1S_{24}, P_2a_3 and P_3a_4 are seminvariants* since each takes the same value for all forms in any class. For the classes (5), (6), (7), their values are S_{24}, a_3 and a_4, respectively. Hence *the five seminvariants* (18) *together with* P_1S_{24}, P_2a_3 *and* P_3a_4 *completely characterize the classes and therefore form a fundamental system of rational integral seminvariants of the quartic form* f *with integral coefficients taken modulo* p, $p > 3$.

Seminvariants of a Modular Binary Form of Order n, §§ 7–13

7. *Fundamental System of Modular Seminvariants Derived by Induction from* $n - 1$ *to* n.—It is necessary to distinguish the case in which the modulus p is prime to n from the case in which p divides n. Binomial coefficients for the form are not permissible in the second case and often not in the first case (for example, if $n = 4$, $p = 3$, since $\binom{n}{2}$ is then divisible by p). Denote the form by

(20) $$F_n = A_0x^n + A_1x^{n-1}y + \cdots + A_ny^n.$$

First, let p be prime to n. For $A_0 \neq 0$, we can transform F_n into a form lacking the second term and having as coefficients the quotients of

(21) $$\sigma_2 = nA_0A_2 - \tfrac{1}{2}(n-1)A_1^2,$$
$$\sigma_3 = n^2A_0^2A_3 - (n-2)nA_0A_1A_2 + \tfrac{1}{3}(n-1)(n-2)A_1^3, \quad \cdots$$

by powers of nA_0. These may also be obtained from (8) by identifying F_n with

(22) $$f_n = a_0x^n + na_1x^{n-1}y + \frac{n(n-1)}{2}a_2x^{n-2}y^2 + \cdots.$$

---

* The first is one-half the discriminant of the semicovariant

$$P_1f/y^2 \equiv P_1(6a_2x^2 + 4a_3xy + a_4y^2) \pmod{p},$$

and the last two are the seminvariants of $P_2f/y^3 \equiv P_2(4a_3x + a_4y) \pmod{p}$.

For p prime to n, a fundamental system of seminvariants of F_n *is given by* $A_0, \sigma_2, \cdots, \sigma_n$ *together with a fundamental system of the particular form of order* $n-1$

$$(23)\qquad \begin{aligned} F'_{n-1} &= P_0F_n/y \\ &\equiv P_0A_1x^{n-1}+P_0A_2x^{n-2}y+\cdots+P_0A_ny^{n-1} \qquad (\text{mod } p), \end{aligned}$$

where $P_0 = 1 - A_0^{p-1}$.

Indeed, $A_0, \sigma_2, \cdots, \sigma_n$ completely characterize the classes of forms F_n with $A_0 \neq 0$. Since $yF_{n-1}' \equiv F_n$ identically, when $A_0 = 0$, the classes of forms F_n with $A_0 = 0$ are completely characterized by the seminvariants of the fundamental system for F_{n-1}'.

For example, A_0 and P_0A_1 form a fundamental system of modular seminvariants of $A_0x + A_1y$ (since these characterize the classes represented by A_0x and A_1y). The corresponding functions for

$$F_1' = P_0A_1x + P_0A_2y$$

are P_0A_1 and

$$\{1 - (P_0A_1)^{p-1}\}P_0A_2 \equiv (1 - A_1^{p-1})P_0A_2 = P_1A_2 \qquad (\text{mod } p).$$

Hence the theorem shows that, *if* $p > 2$,

$$(24)\qquad A_0,\quad 2\sigma_2 = 4A_0A_2 - A_1^2,\quad P_0A_1,\quad P_1A_2$$

form a fundamental system of modular seminvariants of F_2. For f_2, these are

$$(24')\qquad 2a_0,\quad S_2 = a_0a_2 - a_1^2,\quad P_0a_1,\quad P_1a_2.$$

8. *Order a Multiple of the Modulus.*—Next, let $n = pq$. By Fermat's theorem, $x^p - xy^{p-1}$ and hence

$$(25)\qquad \phi = A_0(x^p - xy^{p-1})^q$$

is unaltered modulo p by any transformation (1). Hence if, for each value of the seminvariant A_0, we separate the forms

$$(26)\qquad \overline{F}_{n-1} \equiv \frac{1}{y}(F_n - \phi)$$

into classes under (1), multiply each form by y and add ϕ, we obtain the classes of forms F_n for this value of A_0. Hence, *if n is divisible by p a fundamental system of modular seminvariants of F_n is given by A_0 and a fundamental system for $\overline{F}_{n-1}$.*

For example, if $n = p = 2$,

$$\overline{F}_1 = (A_0 + A_1)x + A_2 y$$

can be transformed into x or $A_2 y$ by (1), according as $A_0 + A_1 \equiv 1$ or 0 (mod 2). Adding $\phi = A_0(x^2 - xy)$ to xy and $A_2 y^2$, we obtain representatives of the classes of forms F_2. Hence the 6 classes are completely characterized by the seminvariants A_0 and those (§ 7) of $\overline{F}_1$, and hence by

$$(27) \qquad A_0, \quad A_1, \quad J \equiv (1 + A_0 + A_1)A_2.$$

9. *Seminvariants of the Binary Cubic Form.*—The classes of algebraic forms f_3 are

$$(28) \qquad a_0 x^3 + 3a_0^{-1}S_2 xy^2 + a_0^{-2}S_3 y^3,$$

$$(29) \qquad 3a_1 x^2 y + \tfrac{1}{4}a_1^{-1}S_{13}y^3, \quad 3a_2 xy^2, \quad a_3 y^3,$$

where the S's are given by (8) and (10_1). The discriminant D of f_3 is given by (13). As in § 3, a_0, S_2, S_3, D form a fundamental system of seminvariants of f_3; they are connected by the syzygy (13).

Henceforth, let the coefficients of f_3 be integers taken modulo p, the excluded case $p = 3$ being treated in § 15. If $p > 3$, the classes are again (28) and (29), and a fundamental system of seminvariants is given by

$$(30) \qquad a_0, \quad S_2, \quad S_3, \quad D, \quad P_1 a_2, \quad P_2 a_3.$$

It is instructive to compare this result with that obtained by the method of § 7. Forming the functions (24) for

$$f_2' = P_0 f_3/y \equiv 3P_0 a_1 x^2 + 3P_0 a_2 xy + P_0 a_3 y^2 \qquad (\text{mod } p),$$

and deleting the factor 3 from the first and second, we get*

$$P_0 a_1, \quad \delta = P_0(4a_1 a_3 - 3a_2^2) = P_0 S_{13}, \quad P_1 a_2, \quad P_2 a_3.$$

* They characterize the classes (29) of f_3 with $a_0 = 0$ and may be so derived.

Hence, if $p > 3$, these four functions and a_0, S_2, S_3 form a fundamental system of modular seminvariants of f_3. We may drop $P_0 a_1$ since

$$(31) \qquad P_0 S_2^{\frac{p-3}{2}} S_3 \equiv \pm 2P_0 a_1^p \equiv \pm 2P_0 a_1 \qquad (\text{mod } p).$$

Hence a fundamental system of seminvariants of f_3 for $p > 3$ is

$$(32) \qquad a_0, \quad S_2, \quad S_3, \quad \delta = P_0 S_{13}, \quad P_1 a_2, \quad P_2 a_3.$$

It is easy to deduce δ from the old set (30), and D from this new set.*

Finally, let $p = 2$. By § 7, a fundamental system of seminvariants for f_3 is given by a_0, S_2, S_3 and a fundamental system for f_2'. The latter system is derived from (27) by replacing A_0, A_1, A_2 by $P_0 a_1$, $P_0 a_2$, $P_0 a_3$, and hence is

$$(1 + a_0)a_1, \quad (1 + a_0)a_2, \quad (1 + a_0)(1 + a_1 + a_2)a_3.$$

We may drop $(1 + a_0)a_1 \equiv (1 + a_0)S_2$.

10. *The Binary Quartic Form.* For $p = 2$, we have

$$\overline{F}_3 = A_1 x^3 + (A_0 + A_2)x^2 y + A_3 x y^2 + A_4 y^3,$$

whose seminvariants are obtained from those of f_3 at the end of § 9. They with A_0 give a fundamental system of seminvariants of F_4:

$$A_0, \quad A_1, \quad A_1 A_3 + A_0 + A_2, \quad (1 + A_1)A_3,$$

$$A_1 A_4 + A_1 A_3 (A_0 + A_2), \quad K = (1 + A_1)(1 + A_0 + A_2 + A_3)A_4.$$

An equivalent fundamental system is†

$$(33) \qquad \begin{gathered} A_0, \quad A_1, \quad A_2 + A_3, \quad (1 + A_1)A_2, \\ A_1 A_4 + A_0 A_2 + A_2 A_3, \quad K. \end{gathered}$$

---

* $D \equiv a_0^{p-3}(S_3^2 + 4S_2^3) - \delta S_2 \qquad (\text{mod } p).$

For, if $a_0 \not\equiv 0$, then $\delta \equiv 0$ and this relation follows from (13); while, if $a_0 = 0$, $D = a_1^2 S_{13} = a_1^2 \delta = - S_2 \delta$. Conversely, δ can be expressed in terms of the functions (30). The above relation gives δS_2. The product of this by S_2^{p-2} is congruent to δ if $S_2 \not\equiv 0$. Also $\delta \equiv 0$ if $a_0 \not\equiv 0$. There remains the case in which $S_2 \equiv 0$, $a_0 \equiv 0$, whence $a_1 \equiv 0$, $\delta \equiv - 3a_2^2 \equiv - 3(P_1 a_2)^2$.

† *Annals of Mathematics*, ser. 2, vol. 15, March, 1914. I there give also a complete set of linearly independent invariants and of linear covariants

For $p > 3$, f_3' is obtained from f_3 by replacing a_0, a_1, a_2, a_3 by

$$4a_1P_0, \quad 2a_2P_0, \quad \tfrac{4}{3}a_3P_0, \quad a_4P_0,$$

respectively. Making this replacement in the second set of seminvariants of f_3 in § 9, we obtain P_0a_1, which may be dropped in view of (31), and the last five functions (34). Hence, *for $p > 3$, a fundamental system of modular seminvariants of f_4 is given by*

$$(34) \quad a_0, \quad S_2, \quad S_3, \quad S_4, \quad P_0S_{13}, \quad P_0S_{14}, \quad P_1S_{24}, \quad P_2a_3, \quad P_3a_4.$$

Here the three S_{ij} are given by (10). Since the functions (34) completely characterize the classes (3)–(7), we have a new proof that they form a fundamental system.

11. *Explicit Fundamental System when $p > n$.*—Instead of employing the above step by step process, we can obtain directly a fundamental system of modular seminvariants of f_n when the modulus p exceeds the order n of the binary form (22). Consider a particular f_n in which a_k is the first non-vanishing coefficient:

$$\sum_{i=k}^{n} \binom{n}{i} a_i x^{n-i} y^i \qquad (a_k \neq 0).$$

To this we apply transformation (1) and obtain

$$\sum_{i=k}^{n} \sum_{j=0}^{n-i} \binom{n}{i} \binom{n-i}{j} a_i t^j x'^{n-i-j} y'^{i+j} = \sum_{l=k}^{n} A_{kl} x'^{n-l} y'^l,$$

where we have replaced j by $l - i$ and set

$$A_{kl} = \sum_{i=k}^{l} \binom{n}{i} \binom{n-i}{l-i} a_i t^{l-i} \equiv \binom{n}{l} \sum_{i=k}^{l} \binom{l}{i} a_i t^{l-i}.$$

Take $k < n$ and give to t the value (2). Thus

$$(35) \qquad A_{kl} = \frac{a_k \binom{n}{l} \sigma_{kl}}{\{(k+1)a_k\}^{l-k}},$$

$$\sigma_{kl} \equiv \sum_{i=k}^{l} (-1)^{l-i} \binom{l}{i} (k+1)^{i-k} a_k^{i-k-1} a_{k+1}^{l-i} a_i.$$

In particular,

$$\sigma_{kk} = 1, \quad \sigma_{kk+1} = 0, \quad \sigma_{0l} = \sum_{i=0}^{l} (-1)^{l-i} \binom{l}{i} a_0^{i-1} a_1^{l-i} a_i,$$

the last being the algebraic seminvariant designated earlier by S_l. It is obtained from the expansion of $(a_0 - a_1)^l$ by replacing a single a_0 in each term by a_i. Except for a numerical factor not divisible by p, σ_{kl} (for $0 < k < l - 1$) equals the S_{kl} in (10) and in (38) below.

The classes C_k of forms f_n in which a_k is the first non-vanishing a are distinguished from each other by the value of a_n if $k = n$, and if $k < n$ by the values of the parameters a_k, σ_{kl} $(l = k + 2, \cdots, n)$. Employing the notation (19), we shall verify that $P_{k-1}a_k$ and $P_{k-1}\sigma_{kl}$ are modular seminvariants of f_n. They vanish for a form C_j $(j \leqq k - 1)$ since then $1 - a_j^{p-1} \equiv 0$. For C_k, they reduce to the parameters a_k and σ_{kl} of that class. For $a_0 = 0, \cdots, a_k = 0$, the first is zero and the second is the expression for σ_{kl} when $a_k = 0$, whose non-vanishing terms (given by $i = k$ and $i = k + 1$) are constant multiples of a_{k+1}^{l-k}; but a_{k+1} is constant for any class C_j $(j > k)$.

It follows also that the parameter a_{k+1} in a class C_{k+1} is determined by the seminvariants $P_{k-1}\sigma_{kl}$ $(l = k + 2,\ k + 3)$, provided $k + 3 \leqq n$. But a_{n-1} and a_n, not so determined, are found from $P_{k-1}a_k$ $(k = n - 1,\ n)$. Hence *a fundamental system of modular seminvariants of f_n, for $p > n$, is given by*

$$\begin{gathered} a_0, \quad \sigma_{0l} \qquad (l = 2, \cdots, n), \\ P_{k-1}\sigma_{kl} \quad (k = 1, \cdots, n-2;\ l = k+2, \cdots, n), \\ P_{n-2}a_{n-1}, \quad P_{n-1}a_n. \end{gathered} \tag{36}$$

For $n = 2, 3, 4$, these are (24′), (32), (34), respectively, except for the difference of notation indicated above. For $n = 5$, we see that *a fundamental system of modular seminvariants of f_5, for $p > 5$, is*

$$\begin{gathered} a_0, \quad S_2, \quad S_3, \quad S_4, \quad S_5, \quad P_0S_{13}, \quad P_0S_{14}, \quad P_0S_{15}, \\ P_1S_{24}, \quad P_1S_{25}, \quad P_2S_{35}, \quad P_3a_4, \quad P_4a_5, \end{gathered} \tag{37}$$

in which the symbols are defined by (8)–(10), (19) *and*

$$(38)\quad \begin{aligned} S_5 &= a_0^4a_5 - 5a_0^3a_1a_4 + 10a_0^2a_1^2a_3 - 10a_0a_1^3a_2 + 4a_1^5,\\ S_{15} &= 16a_1^3a_5 - 40a_1^2a_2a_4 + 40a_1a_2^2a_3 - 15a_2^4,\\ S_{25} &= 27a_2^2a_5 - 45a_2a_3a_4 + 20a_3^3,\\ S_{35} &= 8a_3a_5 - 5a_4^2. \end{aligned}$$

12. *Another Method for the Case $p > n$.*—We may formulate the method of § 7 so that it shall be free from the induction process. The classes of forms (23) with $P_0A_1 \neq 0$, and hence the classes of forms F_n with $A_0 = 0$, $A_1 \neq 0$, are characterized by the seminvariants given by the products of P_0 by the functions $\sigma_2', \cdots$ obtained from $\sigma_2, \sigma_3, \cdots, \sigma_{n-1}$ by increasing the subscript of each A_i by unity and replacing n by $n-1$; indeed, $P_i^2 \equiv P_i \pmod{p}$. When the process of deriving (23) from (20) is applied to (23), we get

$$(39)\quad \begin{aligned} F''_{n-2} &= [1 - (P_0A_1)^{p-1}]F'_{n-1}/y \equiv (1 - A_1^{p-1})P_0F_n/y^2\\ &= P_1F_n/y^2 \equiv P_1A_2x^{n-2} + P_1A_3x^{n-3}y\\ &\qquad + \cdots + P_1A_ny^{n-2} \pmod{p}. \end{aligned}$$

The class of forms (39) with $P_1A_2 \neq 0$, and hence the classes of forms F_n with $A_0 = A_1 = 0$, $A_2 \neq 0$, are characterized by the seminvariants given by the products of P_1 by the functions σ_2'', $\cdots$ obtained from $\sigma_2', \cdots, \sigma_{n-2}'$ by increasing the subscript of each A_i by unity and replacing n by $n-1$. Finally, we obtain $P_{n-2}A_{n-1}x + P_{n-2}A_ny$, characterized by the seminvariants $P_{n-2}A_{n-1}$ and $P_{n-1}A_n$. The earlier $P_{k-1}A_k$ may be dropped (§ 11).

For example, if $n = 3$, $p > 3$, the fundamental system of F_3 is

$$A_0,\quad \sigma_2,\quad \sigma_3,\quad P_0\sigma_2' = P_0(4A_1A_3 - A_2^2),\quad P_1A_2,\quad P_2A_3.$$

Changing the notation from F_3 to f_3, we see that σ_2' becomes $3(4a_1a_3 - 3a_2^2)$, so that the resulting seminvariants are (32). We may of course apply the method directly to f_3; in S_2 we replace a_0, a_1, a_2 by $3a_1, \frac{3}{2}a_2, a_3$ and obtain $\frac{3}{4}(4a_1a_3 - 3a_2^2)$.

Again, to find a fundamental system of f_4 for $p > 3$, we take a_0, S_2, S_3, S_4 and the products of P_0 by the functions $\frac{4}{3}S_{13}$ and $16S_{14}$ obtained from S_2 and S_3 by replacing a_0, a_1, a_2, a_3 by $4a_1$, $\frac{1}{3}\cdot 6a_2$, $\frac{1}{3}\cdot 4a_3$, a_4; then the product of P_1 by the function $2S_{24}$ obtained from S_2 by replacing a_0, a_1, a_2 by $6a_2$, $\frac{1}{2}\cdot 4a_3$, a_4; then P_2a_3 and P_3a_4, to characterize $P_2(4a_3x + a_4y)$. We again have (34).

13. *Number of Linearly Independent Seminvariants.*—Let $p > n$ and employ the notations of § 11. In the classes C_k $(k < n)$. $A_{kk} = a_k\binom{n}{k}$ has $p - 1$ values, $A_{kk+1} = 0$, while $A_{kk+2}, \cdots, A_{kn}$ take independently the values $0, 1, \cdots, p - 1$. In the classes C_n, a_n has p values. Hence there are

$$p + \sum_{k=0}^{n-1}(p-1)p^{n-k-1} = p + p^n - 1$$

distinct classes of forms f_n. Thus by § 11 of Lecture I, *there are exactly $p^n + p - 1$ linearly independent modular seminvariants of f_n when $p > n$.*

Derivation* of Modular Invariants from Seminvariants, §§ 14–15

14. *Invariants of the Binary Quadratic Form.*—First, let $p=2$. Any polynomial in the seminvariants (27) is a linear function of

$$1,\quad A_0,\quad A_1,\quad A_0A_1,\quad J,\quad A_0J \equiv A_0A_1A_2,$$

since $(A_0 + A_1)J \equiv 0$. Since there were six classes, these six seminvariants form a complete set of linearly independent seminvariants. Now a seminvariant is an invariant if and only if it is symmetrical in A_0 and A_2. But

$$I = (1-A_0)(1-A_1)(1-A_2) \equiv (1-A_0)(J+1+A_1) \pmod 2.$$

Thus 1, A_1, A_0J and I are invariants. By subtracting constant

* While this method is usually longer than the method of Lecture I, it requires no artifices and makes no use of the technical theory of numbers. Moreover, it leads to the actual expressions of the invariants in terms of the seminvariants of a fundamental system, thus yielding material of value in the construction of covariants.

multiples of these four, any seminvariant can be reduced to $cA_0 + dA_0A_1$, which is an invariant only when identically zero. Hence 1, A_1, $A_0A_1A_2$ *and* I *form a complete set of linearly independent invariants of* F_2 *modulo* 2.

Next, let $p > 2$. The discriminant of f_2 is $D = S_2$. Any polynomial in the four fundamental seminvariants (24′) is a linear function of

$$a_0^iD^j,\quad P_0a_1^i,\quad P_1a_2^i\quad (i, j = 0, 1, \cdots, p-1),$$

since the product of P_0a_1 or P_1a_2 by a_0 is zero, that of P_1a_2 by P_0a_1 or D is zero, while $DP_0a_1 \equiv -Pa_1^3$. Further,

$$P_0 = 1 - a_0^{p-1},\quad P_0[D^j - (-a_1^2)^j] \equiv 0,$$

$$P_1 = P_0 - P_0a_1^{p-1},\quad a_0^{p-1}D^j \equiv D^j - (-1)^jP_0a_1^{2j},$$

modulo p. Hence any seminvariant is a linear function of

$$(40)\qquad \begin{array}{r} a_0^{p-1},\quad a_0^iD^j\quad (i = 0, 1, \cdots, p-2; j = 0, 1, \cdots, p-1),\\ P_0a_1^k,\quad P_1a_2^k\qquad (k = 1, \cdots, p-1).\end{array}$$

The number of these is $p^2 + p - 1$. Hence (§ 13) *they form a complete set of linearly independent modular seminvariants of* f_2 *for* $p > 2$.

The invariant $A = A_1$ in § 6 of Lecture I becomes for two variables

$$(41)\quad A = \{a_0^\mu + a_2^\mu(1 - a_0^{p-1})\}(1 - D^{p-1}) = a_0^\mu(1 - D^{p-1}) + P_1a_2^\mu,$$

where $\mu = (p-1)/2$. By the expansion of D^{p-1}, we get*

$$(42)\qquad A = (a_0^\mu + a_2^\mu)\left(1 - \sum_{i=0}^{\mu} a_0^ia_2^ia_1^{2\mu-2i}\right).$$

* *Transactions of the American Mathematical Society*, vol. 10 (1909), p. 132. To give a direct proof of the identity of the final expression (41) and (42), note that the product of the final factor in (42) by D equals $a_0a_2 - (a_0a_2)^{\mu+1}$ algebraically, so that the product AD is divisible by p. But the product of (41) by D is evidently divisible by p. It therefore remains only to treat the case $D \equiv 0$. Replacing a_1^2 by a_0a_2, we see that the final factor in (42) becomes $1 - (\mu + 1)a_0^\mu a_2^\mu$. Hence (41) and (42) are now identical if

$$a_0^\mu a_2^\mu(a_0^\mu - a_2^\mu) \equiv 0 \qquad (\text{mod } p).$$

But, if $a_0a_2 \neq 0$, $a_0^\mu a_2^\mu \equiv a_1^{2\mu} \equiv 1$, $a_0^\mu \equiv a_2^\mu \equiv \pm 1$.

Since (42) is therefore a seminvariant and is symmetrical in a_0 and a_2 and since the weight of every term is divisible by $p - 1$, A is an absolute invariant. By (41),

$$(43)\quad \begin{aligned} &A^2 \equiv a_0^{2\mu}(1 - D^{p-1}) + P_1 a_2^{2\mu}, \quad (1 - a_0^{p-1})D^{p-1} \equiv P_0 a_1^{p-1}, \\ &A^2 + D^{p-1} - 1 \equiv -I_0, \quad I_0 = (1-a_0^{p-1})(1-a_1^{p-1})(1-a_2^{p-1}). \end{aligned}$$

Hence also I_0 is an absolute invariant. Subtracting multiples of

$$I_0 = 1 - a_0^{p-1} - P_0 a_1^{p-1} - P_1 a_2^{p-1}, \quad A, \quad D^j \quad (j = 0, 1, \cdots, p-1),$$

we may reduce any seminvariant to a linear function of the expressions (40) other than $P_1 a_2^{p-1}$, $P_1 a_2^{\mu}$, D^j ($j = 0, \cdots, p - 1$). The resulting linear function L is not an invariant. For example, if $p = 3$, it is

$$L = aa_0^2 + ba_0 + ca_0 D + da_0 D^2 + eP_0 a_1 + fP_0 a_1^2 \quad (a, \cdots, f \text{ constants}).$$

Interchange a_0 and a_2, and change the sign of a_1. We get

$$aa_2^2 + ba_2 + ca_2 D + da_2 D^2 + (1 - a_2^2)(fa_1^2 - ea_1).$$

This is to be identically congruent to the invariant L. Taking $a_2 = 0$, we see that $e = f = a = b = 0$, $c = d$. Then $L = ca_0 a_2(a_0 + a_2) + ca_0^2 a_1^2 a_2$ is not symmetric in a_0 and a_2. Hence $L \equiv 0$. For any p, a like result may be proved by considering separately the terms of L of constant weights modulo $p - 1$. Hence in accord with § 11 of Lecture I, *a complete set of linearly independent invariants of f_2, for $p > 2$, is given by I_0, A and the powers of D.* In place of $D^0 = 1$, we may use A^2, in view of (43).

15. *Invariants of the Binary Cubic Modulo* 3.—A fundamental system of seminvariants of F_3 modulo 3 is given by A_0 and a fundamental system of

$$\bar{F}_2 = A_1 x^2 + (A_0 + A_2)xy + A_3 y^2.$$

Hence, by (24), a fundamental system for F_3 is given by

$$A_0, \quad A_1, \quad t = A_1 A_3 - (A_0 + A_2)^2, \quad (1 - A_1^2)(A_0 + A_2),$$
$$\mu = (1 - A_1^2)[1 - (A_0 + A_2)^2]A_3.$$

In place of the fourth and third we may evidently use

$$\lambda = (1 - A_1^2)A_2, \quad \sigma = A_1A_3 + A_0A_2 - A_1^2A_2^2 = t + A_0^2 + \lambda^2.$$

Here σ is the discriminant of F_3 for $p = 3$. By § 13 there are 11 classes of forms $\bar{F}_2$. Hence, by § 8, there are $3 \cdot 11$ classes of forms F_3. Thus there are exactly 33 linearly independent seminvariants of F_3. Since

$$A_1\lambda \equiv A_1\mu \equiv 0, \quad \sigma\lambda \equiv A_0\lambda^2, \quad \mu(\sigma + A_0^2) \equiv 0,$$

$$\mu(\lambda + A_0) \equiv 0, \quad (1 - A_1^2)\sigma \equiv A_0\lambda,$$

modulo 3, any polynomial in the seminvariants A_0, A_1, σ, λ, μ of the fundamental system is congruent to a linear function of

$$(44) \quad A_0^iA_1^j,\ A_0^i\sigma^k,\ A_0^iA_1\sigma^k,\ A_0^i\lambda^k,\ A_0^i\mu^k \quad (i, j = 0, 1, 2;\ k = 1, 2).$$

Hence these 33 functions form a complete set of linearly independent seminvariants of F_3. The seminvariants

$$P = 1 - A_1^2 - \lambda^2 = (1 - A_1^2)(1 - A_2^2),$$

$$(45) \quad I_0 = (1 - A_0^2)(P - \mu^2) = \prod_{i=0}^{3} (1 - A_i^2),$$

$$E = A_0A_1(\sigma - \sigma^2) + A_0\mu = A_0A_3(A_0A_2 - A_1A_3 + A_1^2 - A_2^2)$$

are seen to be invariants as follows.* The weights of the terms of each are all even or all odd. Moreover, under the substitution $(A_0A_3)(A_1A_2)$, induced upon the coefficients of F_3 by the interchange of x and y, the functions σ, P and I_0 are unaltered, while E is changed in sign. Hence σ, P, I_0 are absolute invariants, while E is an invariant of index unity. We now have 7 linearly independent invariants

$$(46) \qquad I_0,\ E,\ E^2,\ \sigma,\ \sigma^2,\ P,\ 1.$$

Noting that

$$(47) \qquad E^2 = A_0^2\mu^2 + A_0^2(\sigma - \sigma^2 + \lambda^2) - A_0\lambda,$$

---

* Or by general theorems, *Transactions of the American Mathematical Society*, vol. 8 (1907), pp. 206–207. Note that E is the eliminant of $F_3 \equiv 0$, $x^3 \equiv x$, $y^3 \equiv y$ (mod 3).

we may employ the functions (46) to delete from (44)

$$\mu^2, \quad A_0\mu, \quad A_0^2\mu^2, \quad \sigma, \quad \sigma^2, \quad \lambda^2, \quad 1$$

in turn (no one of these terms being reintroduced at a later stage). There remain 11 seminvariants of odd weight

$$(48) \qquad A_0^iA_1, \quad A_0^iA_1\sigma, \quad A_0^iA_1\sigma^2, \quad \mu, \quad A_0^2\mu \qquad (i = 0, 1, 2),$$

and 15 of even weight

$$(49) \quad A_0, A_0^2, A_0^iA_1^2, A_0\sigma, A_0\sigma^2, A_0^2\sigma, A_0^2\sigma^2, A_0^i\lambda, A_0\lambda^2, A_0^2\lambda^2, A_0\mu^2.$$

Now the weight and index of a seminvariant of F_3 modulo 3 are both even or both odd.* A linear combination of the functions (48) which is changed in sign by the substitution $(A_0A_3)(A_1A_2)$ is seen to be identically zero (it suffices to set $A_3 = 0$, $A_2 = 0$ in turn). A linear combination of the functions (49) which is unaltered by that substitution is seen similarly to be identically zero. Hence† *a complete set of linearly independent invariants of F_3 modulo 3 is given by* (46).

* When the sign of y is changed, a seminvariant is unaltered or changed in sign according as its weight is even or odd.

† Another proof, using the classes of F_3 under the group of all binary linear transformations of determinant unity modulo 3, and involving a use of more technical theory of numbers, is given in *Transactions of the American Mathematical Society,* vol. 10 (1909), pp. 149–154. The case of any modulus p is there treated.

LECTURE III

INVARIANTS OF A MODULAR GROUP. FORMAL INVARIANTS AND COVARIANTS OF MODULAR FORMS. APPLICATIONS

INVARIANTS OF CERTAIN MODULAR GROUPS, §§ 1–4

1. *Introduction.*—Let G be any given group of g linear homogeneous transformations on the indeterminates $x_1, \cdots, x_m$ with integral coefficients taken modulo p, a prime. Hurwitz* raised the question of the existence of a finite fundamental system of invariants of G. For the relatively unimportant case in which g is not divisible by p, he readily obtained an affirmative answer by use of Hilbert's well known theorem on a set of homogeneous functions, but emphasized the difficulty of the problem in the general case.

In § 5 I shall consider the relation of this question to that of modular covariants and formal invariants of a system of forms and incidentally answer the above question for special groups of orders divisible by p.

I shall, however, first present a simplification of my own work on the total group. Its invariants are universal covariants, i. e., covariants of any system of modular forms (§ 13). It was from the latter standpoint that I was led to the subject of invariants of a modular group independently of Hurwitz's paper, in the title of which the word invariant does not occur.

2. *Invariants of the Total Binary Group.*—Consider the group G of all modular linear homogeneous transformations with integral coefficients of determinant unity:

$$(1) \quad x' \equiv bx + dy, \quad y' \equiv cx + ey, \quad be - cd \equiv 1 \qquad (\text{mod } p).$$

The term point will be used in the sense of homogeneous coordinates, so that $(x, y) = (ax, ay)$, while $(0, 0)$ is excluded.

* *Archiv der Mathematik und Physik*, (3), vol. 5 (1903), p. 25.

We do not restrict the coordinates to be integers, but permit their ratio to be a root of any congruence with integral coefficients modulo p. A point is called *real* if the ratio of its coordinates is rational.

A point (x, y) is invariant under a transformation (1) if $x' \equiv \rho x$, $y' \equiv \rho y$, or

$$(2) \quad (b - \rho)x + dy \equiv 0, \quad cx + (e - \rho)y \equiv 0 \qquad (\mathrm{mod}\ p).$$

If these congruences hold identically as to x, y, then

$$d \equiv c \equiv 0, \quad b \equiv e \equiv \pm 1 \qquad (\mathrm{mod}\ p)$$

and the transformation is one of the transformations

$$(3) \qquad x' \equiv \pm x, \quad y' \equiv \pm y \qquad (\mathrm{mod}\ p),$$

which leave every point invariant.

A *special* point is one invariant under at least one transformation (1) not of the form (3). There are $p(p^2 - 1)$ transformations (1). We shall assume in the text that $p > 2$ (relegating to foot-notes the modifications to be made when $p = 2$). Then there are two transformations (3). Hence any non-special point is one of exactly*

$$(4) \qquad \omega = \tfrac{1}{2}p(p^2 - 1)$$

conjugate points under the group G, while a special point is one of fewer than ω conjugates.

Let (x, y) be a special point and let (1) be a transformation, not of the form (3), which leaves it invariant. Thus the congruences (2) are not both identities. The determinant of their coefficients must therefore be divisible by p. Hence ρ is a root of the *characteristic* congruence (in which $\alpha = b + e$)

$$(5) \qquad \rho^2 - \alpha\rho + 1 \equiv 0 \qquad (\mathrm{mod}\ p).$$

First, suppose that (5) has an integral root ρ. For this value of ρ, one of the congruences (2) is a consequence of the other, and the ratio $x : y$ is uniquely determined as an integer modulo p.

* For $p = 2$, ω is to be replaced by $2(2^2 - 1) = 6$.

Hence only real special points are invariant under a transformation [other than (3)] whose characteristic congruence has an integral root. Moreover, all real points are conjugate under the group G. Indeed,

$$x' \equiv bx, \quad y' \equiv x + b^{-1}y, \quad \text{and} \quad x' \equiv -y, \quad y' \equiv x$$

replace (1, 0) by $(b, 1)$ and (0, 1) respectively. Hence if an invariant of G vanishes for one of the real points, it vanishes for all and has the factor

$$L = y\prod_{a=0}^{p-1}(x - ay) \equiv x^p y - xy^p \pmod{p}, \tag{6}$$

the congruence following from Fermat's theorem. Obviously, any transformation of G replaces a real point by a real point, and therefore L by kL. The constant k is in fact unity and L is an invariant of G. Indeed, for

$$x \equiv aX + bY, \quad y \equiv cX + dY \pmod{p}, \tag{7}$$

where $a, \cdots, d$ are integers of determinant $\Delta = ad - bc$,

$$\begin{vmatrix} x^p & y^p \\ x & y \end{vmatrix} \equiv \begin{vmatrix} aX^p + bY^p & cX^p + dY^p \\ aX + bY & cX + dY \end{vmatrix} = \Delta \begin{vmatrix} X^p & Y^p \\ X & Y \end{vmatrix} \pmod{p}. \tag{8}$$

Next, suppose that (5) has no integral root and therefore two Galois imaginary roots. By (2), each root ρ uniquely determines a point (x, y) with $y \not\equiv 0$. We may therefore take $y = 1$, whence $cx \equiv \rho - e$. The resulting two special points are therefore imaginary points of the form $(r\rho + s, 1)$, where r and s are integers modulo p, and r is not divisible by p. The imaginaries introduced* by new transformations are expressible linearly in terms of this ρ. Indeed, $(2\rho - \alpha)^2 \equiv A$, where $A = \alpha^2 - 4$ is a quadratic non-residue of p (i. e., is not the remainder when the square of any integer is divided by p). Thus $A \equiv a^2\nu$, where ν is a fixed non-residue of p. Hence the roots of all congruences (5) having no integral roots are expressible in the form $k + l\sqrt{\nu}$, where k and l are integers.

* There are no new ones if $p = 2$, since $\alpha \equiv 0 \pmod{2}$.

Hence the special points invariant under transformations whose characteristic congruences have no integral roots are all of the form $(r\rho + s, 1)$, where r and s are integers, r not divisible by p, while ρ is a fixed root of a particular one of these congruences (5).

We next show that these $p^2 - p$ imaginary special points are all conjugate under the group G. It suffices to prove that they are all conjugate with $(\rho, 1)$, which is invariant under

$$x' \equiv \alpha x - y, \quad y' \equiv x.$$

Now transformation (1) replaces $(\rho, 1)$ by $(R, 1)$, where

$$R = \frac{b\rho + d}{c\rho + e}.$$

We are to prove that there exist integers b, c, d, e satisfying

$$(9) \qquad be - cd \equiv 1 \pmod{p},$$

such that $R \equiv r\rho + s$, where r and s are any assigned integers for which r is not divisible by p. Denote the second root of (5) by ρ' and multiply the numerator and denominator of R by $c\rho' + e$. Using (9), we get

$$R \equiv \frac{\rho + N}{q}, \quad N = bc + de + dc\alpha, \quad q = c^2 + \alpha ce + e^2.$$

We first show* that we can choose integers c and e such that $q \equiv i \pmod{p}$, where i is any assigned integer not divisible by p. If i is a quadratic residue of p, we may take $c = 0$. Next, let i be a quadratic non-residue of p. Taking $c \not\equiv 0$, $e \equiv kc$, we have

$$q \equiv c^2 f(k), \quad f(k) = 1 + \alpha k + k^2.$$

Now $f(k) \equiv f(K)$ if and only if $K \equiv k$ or $K \equiv -\alpha - k$. Hence the $p - 1$ values of k other than $-\alpha/2$ give by pairs the same value of $f(k)$. Thus for $k = 0, \cdots, p-1$, $f(k)$ takes $1 + \frac{1}{2}(p-1)$ incongruent values, no one a multiple of p [since (5) has no

* If $p = 2$, then $\alpha \equiv 0$; taking $c = 1$, $e = 0$, we have $q \equiv 1 \equiv i \pmod{2}$.

integral root], and consequently a value which is a quadratic non-residue of p. Then, by choice of c, q can be made congruent to any assigned non-residue.

Having made $q \equiv i \pmod{p}$ by choice of c and e, we proceed to choose integral solutions b and d of (9) such that N will be congruent to any assigned integer j. If $c \equiv 0$, so that $e \not\equiv 0$, we take $d \equiv j/e$. If $c \not\equiv 0$, we eliminate d from N by use of (9) and obtain

$$N \equiv \frac{1}{c}(bq - e - c\alpha), \quad q = c^2 + \alpha ce + e^2.$$

Since $q \not\equiv 0$, we may make $N \equiv j$ by choice of b.

We have therefore proved that there are exactly $p^2 - p$ imaginary special points, viz., $(r\rho + s, 1)$, $r \not\equiv 0$, and that they are all conjugate under the group G. Hence any invariant of G which vanishes for an imaginary special point has the factor

$$(10) \qquad Q = \frac{x^{p^2}y - xy^{p^2}}{L} = \frac{x^{p^2-1} - y^{p^2-1}}{x^{p-1} - y^{p-1}}.$$

Indeed, the numerator of the first fraction vanishes for $x = r\rho + s$, $y = 1$, since

$$(r\rho + s)^{p^2} \equiv r\rho^{p^2} + s, \quad \rho^{p^2} \equiv \rho \pmod{p},$$

the last congruence* being a case of Galois's generalization of Fermat's theorem. We have divided out L, which vanishes for the real points $(s, 1)$ and $(1, 0)$. Since any transformation of G replaces one of our imaginary points by another, it replaces Q by kQ. The constant k is in fact unity and Q is an invariant of G. Indeed, (8) holds if we replace the exponents p by p^2. Hence the quotient Q is invariant† under all transformations (7).

* It may be proved by noting that (5) implies

$$(\rho^2 - \alpha\rho + 1)^p \equiv \rho^{2p} - \alpha\rho^p + 1 \equiv 0 \pmod{p},$$

so that ρ^p is the second root of (5). By the same argument, $(\rho^p)^p$ is a root, distinct from ρ^p, and hence identical with ρ.

† I gave the notation Q to the invariant (10) since it is the product of all of the binary quadratic forms $x^2 + \cdots$ which are irreducible modulo p. Indeed, the latter vanishes for two points of the form $(r\rho + s, 1)$ and $(r\rho' + s, 1)$, where ρ and ρ' are the roots of (5) and r, s are integers, $r \not\equiv 0$, and conversely.

We are now ready to prove that *any rational integral invariant I, with integral coefficients, of the group G is a rational integral function of L and Q with integral coefficients.*

After removing possible factors L and Q, we may assume that I vanishes for no special point. If I is not a constant, it vanishes at a point (c, d) and hence at the ω distinct points conjugate with (c, d) under the group G. The invariants*

$$(11) \qquad q = Q^{\frac{p+1}{2}}, \quad l = L^{\frac{p(p-1)}{2}}$$

are of degree ω. The constant τ, determined by

$$q(c, d) + \tau \cdot l(c, d) \equiv 0 \qquad (\text{mod } p),$$

is a root of a congruence of a certain degree t with integral coefficients and irreducible modulo p. Now $q + \tau l$ is a factor of I. Since q, l and I have integral coefficients, I has also the factors

$$(12) \qquad q + \tau^p l, \quad q + \tau^{p^2} l, \quad \cdots, \quad q + \tau^{p^{t-1}} l.$$

For, by Galois's theorem mentioned above,

$$\tau, \quad \tau^p, \quad \tau^{p^2}, \quad \cdots, \quad \tau^{p^{t-1}}$$

are the roots of our irreducible congruence of degree t. Since the conditions which imply that $q + zl$ shall be a factor of I are congruences satisfied when $z = \tau$, they are satisfied when $z = \tau^{p^k}$. Hence if we multiply $q + \tau l$ by the product of the invariants (12), we obtain an invariant T with integral coefficients modulo p. Since L and Q have no common factor, no two of the functions $q + \tau l$ and (12) have a common factor. Hence T is a factor of I. Proceeding in like manner with I/T, we arrive finally at the truth of the theorem.†

3. *Invariants of Smaller Binary Groups.*—We shall later need the theorem that *a fundamental system of rational integral invariants*

* If $p = 2$, we omit the divisor 2 in the exponents.

† Proved less simply in *Transactions of the American Mathematical Society*, vol. 12 (1911), p. 1. Still simpler is the proof that various coefficients of an invariant are zero, *Quarterly Journal of Mathematics*, 1911, p. 158.

of the group composed of the p powers of the transformation

$$(13)\qquad x' \equiv x + y, \quad y' \equiv y \qquad (\text{mod } p)$$

is given by y and λ, where

$$(14)\quad \lambda = x(x+y)(x+2y)\cdots(x+\overline{p-1}y) \equiv x^p - xy^{p-1} \qquad (\text{mod } p).$$

Now (1, 0) is the only special point, being the only point unaltered by (13) or its kth power, $k < p$. Hence an invariant not having a factor y or λ vanishes at imaginary points falling into sets each of p points conjugate under our group. As at the end of § 2, the invariant is a product of factors $y^p + \tau\lambda$ so related that the product equals a polynomial in y^p and λ with integral coefficients.

Other results will be merely stated, since they are not presupposed in what follows. Within the group G of all transformations (1), any subgroup of order a multiple of p is conjugate with one containing (13) and transformations exclusively of the form

$$(15)\qquad x' \equiv tx + ly, \quad y' \equiv t^{-1}y \qquad (\text{mod } p),$$

and having y and λ as a fundamental system of invariants.* The invariants of any subgroup whose order is prime to p have been found.†

4. *Invariants of the Total Group on m Variables.*—The functions

$$(16)\quad L_m = \begin{vmatrix} x_1^{p^{m-1}} & \cdots & x_m^{p^{m-1}} \\ x_1^{p^{m-2}} & \cdots & x_m^{p^{m-2}} \\ \cdot & \cdots & \cdot \\ x_1^{p} & \cdots & x_m^{p} \\ x_1 & \cdots & x_m \end{vmatrix}, \quad Q_{ms} = \begin{vmatrix} x_1^{p^m} & \cdots & x_m^{p^m} \\ \cdot & \cdots & \cdot \\ x_1^{p^{s+1}} & \cdots & x_m^{p^{s+1}} \\ x_1^{p^{s-1}} & \cdots & x_m^{p^{s-1}} \\ \cdot & \cdots & \cdot \\ x & \cdots & x_m \end{vmatrix} \div L_m$$

are seen, by a generalization of (8), to be invariants of index 1 and 0 respectively of the group Γ_m of all linear homogeneous transformations on $x_1, \cdots, x_m$ with integral coefficients modulo p.

* *Bulletin of the American Mathematical Society*, vol. 20 (1913), pp. 132–4.
† *American Journal of Mathematics*, vol. 33 (1911), p. 175.

Since L_m is an invariant of Γ_m and has the factor x_1, it follows from an examination of its diagonal term that*

$$(17)\quad L_m \equiv \prod_{k=1}^{m} \sum_{c_i=0}^{p-1} (x_k + c_{k+1}x_{k+1} + \cdots + c_m x_m) \qquad (\text{mod } p),$$

in which occurs one of each set of proportional linear forms modulo p. A like proof shows that the numerator of Q_{ms} is divisible by each of the linear functions (17) and hence by L_m, modulo p.

Making use of the theorem in § 2, I have proved by induction† that the m invariants $L_m, Q_{m1}, \cdots, Q_{mm-1}$ are independent and form a fundamental system of rational integral invariants of Γ_m.

A fundamental system of invariants of the group of all modular linear transformations on two sets of two cogredient variables has been obtained very recently by Dr. W. C. Krathwohl in his Chicago dissertation.‡

Formal Invariants and Seminvariants of Modular Forms, §§ 5–13

5. *Formal Modular Invariants.*—Consider a binary form

$$f(x, y) = a_0x^r + a_1x^{r-1}y + \cdots + a_ry^r,$$

in which $x, y, a_0, \cdots, a_r$ are arbitrary variables. The transformation (7) with integral coefficients, whose determinant Δ is not divisible by the prime p, replaces f by a form

$$\phi(X, Y) = A_0X^r + A_1X^{r-1}Y + \cdots + A_rY^r,$$

in which

$$(18)\quad A_0 = f(a, c),\quad A_1 = ra^{r-1}ba_0 + \cdots,\quad \cdots,\quad A_r = f(b, d).$$

A polynomial $P(a_0, \cdots, a_r)$ with integral coefficients is called a formal invariant modulo p of index λ of f under the transforma-

* E. H. Moore, *Bulletin of the American Mathematical Society*, vol. 2 (1896), p. 189. His proofs do not use the invariantive property. A like remark is true of the proof that the product (17), in the case $x_m = 1$, is congruent to a determinant of order $m - 1$, then obviously equal to L_m, by R. Levavasseur, *Mémoires de l'Académie des Sciences de Toulouse*, ser. 10, vol. 3 (1903), pp. 39–48; *Comptes Rendus*, 135 (1902), p. 949.

† *Transactions of the American Mathematical Society*, vol. 12 (1911), p. 75.

‡ *American Journal of Mathematics*, October, 1914.

tion (7) if

$$(19) \qquad P(A_0, A_1, \cdots, A_r) \equiv \Delta^\lambda P(a_0, a_1, \cdots, a_r) \pmod{p},$$

identically as to $a_0, \cdots, a_r$, after the A's have been replaced by their values (18) in terms of the a_i. If P is invariant modulo p under all transformations (7), it is called a formal invariant modulo p of f.

The term formal is here used in connection with a form f whose coefficients are arbitrary variables in contrast to the case, treated in the earlier Lectures, in which the coefficients are undetermined integers taken modulo p. In the latter case, (19) necessarily becomes an identical congruence in the a's only after the exponent of each a is reduced to a value less than p by means of Fermat's theorem $a^p \equiv a \pmod{p}$.

The functions (18) are linear in $a_0, \cdots, a_r$. It is customary to say that relations (18) define a linear transformation on $a_0, \cdots, a_r$ which is induced by the binary transformation (7). Let Γ be the group of all of the transformations (18) induced by the group of all of the binary transformations (7). Making no further use of the form f, we may state the above problem of the determination of the formal invariants of f in the following terms. We desire a fundamental system of invariants of group Γ. This problem is of the type proposed in § 1; the group Γ is a special group of order a multiple of p. Here and below the term invariant is restricted to rational integral functions of $a_0, \cdots, a_r$.

A theory of formal invariants has not been found. For no form f has a fundamental system of formal invariants been published. Some light is thrown upon this interesting but difficult problem by the following complete treatment of a binary quadratic form, first for the exceptional case $p = 2$ and next for the case $p > 2$, and preliminary treatment of a binary cubic form.

6. *Formal Invariants Modulo 2 of a Binary Quadratic Form.*—Write

$$(20) \qquad f = ax^2 + bxy + cy^2,$$

where a, b, c are arbitrary variables. Under the transformation

$$(21) \qquad x = x' + y', \quad y = y',$$

f becomes f', in which the coefficients are

$$(22) \qquad a' \equiv a, \quad b' \equiv b, \quad c' \equiv a + b + c \qquad (\text{mod } 2).$$

By § 3, the only invariants under $d' \equiv d$, $c' \equiv c + d$, modulo 2, are the polynomials in d and $c(c + d)$. Take $d = a + b$. Hence *the only seminvariants of f are the polynomials in a, b and*

$$(23) \qquad s = c(c + a + b).$$

Such a polynomial is an invariant of f if and only if it is unaltered by the substitution (ac) induced by (xy). Thus

$$(24) \quad b, \quad k = as, \quad q = b(a + c) + a^2 + ac + c^2 = s + ab + a^2$$

are invariants of f. Introducing q in place of s, we see that any seminvariant is a polynomial in a, b, q. Consider an invariant of this type. Since its terms free of a are invariants, the sum of its terms involving a is an invariant with the factor a and hence also the factors c and $a + b + c$, the last by (22). Hence this sum has the factor k, and its quotient by k is an invariant. By induction we have the theorem:

Any rational integral formal invariant of f equals a rational integral function of b, q, k.*

7. *Formal Seminvariants of a Binary Quadratic Form for $p > 2$.* Write

$$(25) \qquad f = ax^2 + 2bxy + cy^2,$$

where a, b, c are arbitrary variables. Under the transformation (21), f becomes f', whose coefficients are

$$(26) \qquad a' = a, \quad b' = a + b, \quad c' = a + 2b + c.$$

* Replace x_1, x_2, x_3, of § 4 by a, b, c; then

$$L_3 = bk(k + bq), \quad Q_{32} = b^4 + bk + q^2, \quad Q_{31} = b^2q^2 + bqk + b^3k + k^2.$$

Evident formal seminvariants are a, $\Delta = b^2 - ac$, and

$$(27) \qquad \beta = \prod_{t=0}^{p-1} (ta + b) \equiv b^p - ba^{p-1} \qquad (\text{mod } p),$$

$$(28) \qquad \gamma_k = \prod_{t=0}^{p-1} \{(t^2 - k)a + 2tb + c\} \quad (k = 0, 1, \cdots, p - 1).$$

Indeed, the linear function under the product sign in (28) is transformed by (26) into the function derived from it by replacing t by $t + 1$. As in (27),

$$(29) \qquad [\gamma_k]_{a=0} \equiv c^p - cb^{p-1} \qquad (\text{mod } p).$$

Let $S(a, b, c)$ be a homogeneous rational integral seminvariant with integral coefficients. Then, by (26),

$$S(0, b, c) \equiv S(0, b, 2b + c) \qquad (\text{mod } p).$$

Thus, by § 3, $S(0, b, c)$ equals a polynomial in b, $c^p - cb^{p-1}$. Hence, by (29),

$$S(a, b, c) \equiv a\sigma(a, b, c) + \phi(b, \gamma_k) \qquad (\text{mod } p),$$

where σ and ϕ are polynomials in their arguments. Now

$$b^{2i} = \Delta^i + a(\quad), \quad b^{p+2i} = \beta\Delta^i + a(\quad).$$

Hence

$$(30) \qquad S = a\lambda(a, b, c) + \psi(\beta, \Delta, \gamma_k) + \sum_{i=0}^{(p-3)/2} d_i b^{2i+1} \gamma_k{}^{e_i},$$

where λ and ψ are polynomials in their arguments, and d_i is an integer.

When y is multiplied by a primitive root ρ of p, a, b, c are multiplied by 1, ρ, ρ^2, respectively. Hence β is multiplied by ρ, while, by (29), γ_k and Δ are multiplied by ρ^2. If therefore we attribute the weights 0, 1, 2 to a, b, c, respectively, and the weight $s + 2t$ to $a^r b^s c^t$, we see that the weight of every term of γ_k is congruent to 2 modulo $p - 1$.

We can now prove that every d_i is divisible by p. For, if not, the seminvariant $S - \psi$ has a term of odd weight, so that every term of λ is of odd weight and hence has the factor b. Thus $S - \psi$ has the factor b and therefore the factor β, so that its terms free of a have the factor b^p. But this is impossible, since $2i + 1 < p$ and (29) does not have the factor b.

Hence $S - \psi$ has the factor a and the quotient is a seminvariant of the form $a\lambda' + \psi'$. Proceeding in this way, we obtain the theorem:

Any seminvariant is a polynomial in a, Δ, β and any single γ_k.

Of these, β alone is of odd weight. Hence any seminvariant is a polynomial in $a, \Delta, \gamma_k, \beta^2$ or the product of such a polynomial by β. But

$$(31) \qquad \beta^2 \equiv a^p\gamma_0 + \Delta(\Delta^{\frac{p-1}{2}} - a^{p-1})^2 \qquad (\text{mod } p).$$

To prove this, it suffices to show that the second member is divisible by b and hence by β, and being of even weight therefore by β^2, and to remark that each member of (31) reduces to b^{2p} for $a = 0$. Now

$$[\gamma_0]_{b=0} = \prod_{t=0}^{p-1}(t^2a + c) = c\left\{\prod_{t=1}^{(p-1)/2}(t^2a + c)\right\}^2$$

$$\equiv c\{c^{\frac{p-1}{2}} - (-a)^{\frac{p-1}{2}}\}^2 \qquad (\text{mod } p),$$

$$a^p[\gamma_0]_{b=0} \equiv ac\{(-ac)^{\frac{p-1}{2}} - a^{p-1}\}^2 \qquad (\text{mod } p).$$

But Δ reduces to $-ac$ for $b = 0$. Hence the second member of (31) has the factor b. We therefore have the theorem:

For $p > 2$, any formal seminvariant of a binary quadratic form is a polynomial in a, Δ, γ_0 or the product of such a polynomial by β.

8. *Formal Invariants of a Binary Quadratic Form for $p > 2$.* The product

$$(32) \quad \Gamma = \prod_k \gamma_k \quad (k \text{ ranging over the quadratic non-residues of } p)$$

is an absolute invariant of f under the group G of all binary transformations with integral coefficients taken modulo p of determinant unity. It suffices to prove that this seminvariant is unaltered by the substitution

$$a' = c, \quad c' = a, \quad b' = -b, \tag{33}$$

induced by the transformation $x = y'$, $y = -x'$. Under (33), the general factor in (28) is replaced by

$$(t^2 - k)\{(T^2 - K)a + 2Tb + c\},$$

where

$$T = \frac{-t}{t^2 - k}, \qquad K = \frac{k}{(t^2 - k)^2}.$$

Hence K is quadratic non-residue of p when k is. Also,

$$\prod_{t=0}^{p-1}(t^2-k) = -k\left\{\prod_{t=1}^{(p-1)/2}(k-t^2)\right\}^2 \equiv -k(k^{\frac{p-1}{2}}-1)^2 \equiv -4k \pmod{p}$$

if k is a non-residue. To show that the product of the resulting numbers $-4k$ is congruent to unity, we set $x = 0$ in

$$\prod_k (x - k) \equiv x^{\frac{p-1}{2}} + 1 \pmod{p}, \tag{34}$$

and note that $2^{p-1} \equiv 1$. Hence (32) is unaltered by (33) and is an absolute invariant of f under G.

It is very easy to verify that

$$J = a\gamma_0 \tag{35}$$

is unaltered by (33), so that J is an invariant of f under G.

If an invariant has the factor β, it has the factor

(36) $B = \beta \Pi \gamma_r$ (r ranging over the quadratic residues of p).

For, under the substitution (33), $b + \tau a$ ($\tau \neq 0$) becomes $\tau(c - b/\tau)$. By choice of τ, we reach $c + 2tb$, where t is any assigned integer not divisible by p. This is a factor of γ_k where $k \equiv t^2$.

The fact that B is an invariant may be verified as in the case of (32) or deduced from the fact that

$$a\beta \prod_{k=0}^{p-1} \gamma_k = a\gamma_0 \cdot B\Gamma$$

is an invariant, being the product of all non-proportional linear functions of a, b, c with integral coefficients modulo p.

Hence any invariant is the product of a power of B by an invariant which is a polynomial P in a, Δ, γ_0.

Since γ_k is a seminvariant not divisible by β, it equals a polynomial in a, Δ, γ_0 (§ 7). But if $a = 0$, $\gamma_k \equiv \gamma_0 \pmod{p}$, by (29), and $\Delta = b^2$ is free of c, so that γ_k is not a polynomial in a and Δ only. Hence

$$\gamma_k \equiv \gamma_0 + g_k(a, \Delta) \pmod{p}. \tag{37}$$

For $p = 3$, the polynomial P therefore equals a polynomial in a, Δ, $\gamma_2 = \Gamma$. Now an invariant $\phi(a, \Delta, \Gamma)$ differs from the invariant $\phi(0, \Delta, \Gamma)$ by an invariant with the factor a and hence the factor (35). Treating the quotient similarly, we ultimately obtain the following theorem for the case $p = 3$:

A fundamental system of formal invariants of the binary quadratic form f modulo p, $p > 2$, is given by the discriminant Δ and Γ, J, B, defined by (32), (35), (36). *The product of the last three is congruent modulo p to the product of all the non-proportional linear functions of the coefficients of f.*

To prove the theorem for $p > 3$, note first, by (37), that Γ, given by (32), differs from γ_0^n by a polynomial in γ_0, a, Δ of degree $n - 1$ in γ_0, where $n = (p - 1)/2$. Hence a polynomial in a, Δ, γ_0 equals a polynomial in a, Δ, γ_0, Γ of degree at most $n - 1$ in γ_0. Subtract from each the terms of the latter involving only the invariants Δ, Γ. We have therefore to investigate invariants of the type

$$\sum_{i=1}^{n-1} c_i \gamma_0^i P_i(\Delta, \Gamma) + \sum_{i=0}^{n-1} \gamma_0^i \phi_i(a, \Delta, \Gamma), \tag{38}$$

in which the c_i are integers, while P_i and ϕ_i are polynomials in their arguments, and ϕ_i has the factor a. If every $c_i \equiv 0$, the invariant has the factor a and hence the factor $a\gamma_0 = J$, and the quotient by J is an invariant which may be treated similarly. The theorem will therefore follow if we show that a contradiction is involved in the assumption that a certain c_j is not divisible by p. First, the remaining c_i are divisible by p. For if also $c_i \not\equiv 0$, let $k_i\Delta^{r_i}\Gamma^{s_i}$ be the term of P_i of highest degree in Δ. Since γ_0 and Γ are of degrees p and np, and of weights $\equiv 2$ and $0 \pmod{p-1}$, $\gamma_0{}^iP_i$ is of degree $pi + 2r_i + s_inp$ and of weight $\equiv 2i + 2r_i \pmod{p-1}$. But $p \equiv 1 \pmod{n}$. Hence

$$i + 2r_i \equiv j + 2r_j, \quad 2i + 2r_i \equiv 2j + 2r_j \qquad (\text{mod } n),$$

so that $i \equiv j \pmod{n}$. But i and j are positive integers $< n$. Hence $i = j$. Multiplying our invariant by a suitably chosen integer, we have the invariant

$$(39) \quad \gamma_0{}^jP_j(\Delta, \Gamma) + \sum_{i=0}^{n-1}\gamma_0{}^i\phi_i(a, \Delta, \Gamma), \quad P_j = \Delta^r\Gamma^s + \cdots.$$

Now $-(c - ka)b^{p-1}$ is the term of highest degree in b in γ_k. Hence

$$(40) \qquad \gamma_0 = -cb^{p-1} + \cdots, \quad \Gamma = \sigma b^{n(p-1)} + \cdots,$$

$$(41) \quad \sigma = \prod_k\{-(c - ka)\} \equiv (-c)^n + (-a)^n \qquad (\text{mod } p),$$

where k ranges over the non-residues of p, the last following from (34) for $x = c/a$. Since γ_0 and Γ are of even weights, only even powers of b enter (39). Hence an invariant (39) is symmetrical in a and c. We shall prove that this is not the case for the terms of highest degree in b. For $\gamma_0{}^jP_j$ this term is

$$(42) \qquad (-c)^j\sigma^sb^\beta, \quad \beta = j(p-1) + 2r + sn(p-1).$$

Let $C_ia^{e_i}\Delta^{f_i}\Gamma^{g_i}$ be one of the terms of ϕ_i in which the exponent of b is a maximum. Then in $\gamma_0{}^i\phi_i$ the highest power of b occurs in the terms

$$(43) \quad C_ia^{e_i}(-c)^i\sigma^{g_i}b^{\beta_i}, \quad \beta_i = 2f_i + g_in(p-1) + i(p-1).$$

Since the weight and degree is the same as for (42),

$$(44) \qquad \begin{aligned} 2i + \beta_i &\equiv 2j + \beta \qquad (\text{mod } p-1), \\ e_i + i + g_in + \beta_i &= j + sn + \beta. \end{aligned}$$

First, let $\beta_i = \beta$. Then $i \equiv j$, $e_i \equiv 0 \pmod{n}$, whence $i = j$. Thus the exponent of a in any term (42) or (43) is divisible by n, while the exponent of c is not, being congruent to j modulo n. Hence the coefficient of b^β in the sum of (42) and the various terms (43), with $i = j$, is not symmetrical in a and c, unless identically zero. But (43) has the factor a while (42) does not. Hence the greatest β_i exceeds β.

Next, consider a set of terms (43) and a set of terms of like form with i replaced by k, all being of equal degree in b. Then $\beta_i = \beta_k$. By (44_1), $2i + \beta_i \equiv 2k + \beta_k$, $i = k$. Consider finally terms (43) with β_i constant. In them the residue modulo n of e_i is a constant $\neq i$. For, if $e_i \equiv i$, then $2i + \beta_i \equiv j + \beta \pmod{n}$ by (44_2), so that $j \equiv 0 \pmod{n}$ by (44_1). Hence these terms (43) are not symmetric in a and c and yet do not cancel.*

Our fundamental invariants are connected by a syzygy; for $p = 3$,

$$B^2 \equiv \Delta^3\Gamma^2 + J(J - \Delta^2)^2. \tag{45}$$

9. *Formal Invariants of a Binary Cubic Form for* $p \neq 3$.— We have seen that the theory of formal invariants of a binary quadratic form is dominated by the invariantive products of linear functions of the coefficients. While these products depended upon the classification of integers into the quadratic residues and the non-residues of p, we shall find that for a cubic form it is a question not merely of cubic residues and non-residues of p, but of the larger classes of reducible and irreducible congruences. Write

$$f = ax^3 + 3bx^2y + 3cxy^2 + dy^3,$$

thus taking $p \neq 3$. Under transformation (21), f becomes f', whose coefficients are given by (26) and

$$d' = a + 3b + 3c + d. \tag{46}$$

---

* If two are of like degree in c, their g's are equal and hence their f's are equal; then, if of like degree in a, their e's are equal. But then we have the same term of ϕ_i.

Hence a, β and γ_k, given by (27) and (28), are again seminvariants; also,

$$(47)\quad \delta_{jk} = \prod_{t=0}^{p-1} \{(t^3 - 3kt - j)a + 3(t^2 - k)b + 3tc + d\} \qquad (j, k = 0, \cdots, p-1).$$

Indeed, if $F_t(a, b, c, d)$ is the function in brackets,

$$F_t(a', b', c', d') = F_{t+1}(a, b, c, d).$$

Any invariant with the factor a has the factor

$$(48)\quad a\delta_{00} = a \prod_{t=0}^{p-1} (t^3a + 3t^2b + 3tc + d) = f(1, 0) \prod_{t=0}^{p-1} f(t, 1),$$

whose vanishing is the condition that one of the points (x, y) represented by $f = 0$ shall be one of the existing $p + 1$ real points $(1, 0)$, $(t, 1)$ of the modular line. To verify algebraically that the seminvariant (48) is an invariant,* note that it is unaltered modulo p by the substitution

$$(49)\qquad a' = -d, \quad d' = a, \quad b' = c, \quad c' = -b,$$

which is induced on the coefficients of f by $x = y'$, $y = -x'$.

The product P of the δ_{jk} in which j and k are such that

$$\lambda = t^3 - 3kt - j$$

is irreducible modulo p is a formal invariant.

The substitution (49) replaces the general factor of (47) by

$$-a + 3tb - 3(t^2 - k)c + \lambda d$$
$$= \lambda\{(T^3 - 3KT - J)a + 3(T^2 - K)b + 3Tc + d\},$$

where

$$T = \frac{k - t^2}{\lambda}, \qquad K = \frac{g}{\lambda^2}, \qquad J = \frac{h}{\lambda^3}, \qquad g = k^2 + kt^2 + tj,$$

$$h = -2k^3 + 6k^2t^2 + 3ktj + t^3j + j^2.$$

---

* For any form, see *Transactions of the American Mathematical Society*, vol. 8 (1907), pp. 207–208.

We are to show that there is no integral solution x of

$$x^3 - 3Kx - J \equiv 0 \qquad (\text{mod } p).$$

Multiply this by λ^3 and set $\lambda x = y$. Then

$$y^3 - 3gy - h \equiv 0 \qquad (\text{mod } p).$$

But the negative of the left member is the result of substituting

$$r + s = -t, \quad rs = -y - 2k$$

in the expansion of the product

$$(r^3 - 3kr - j)(s^3 - 3ks - j).$$

The latter is congruent to zero modulo p for no values of r and s which are integers or the roots of an irreducible quadratic congruence with the integral coefficients t, $-y - 2k$.

For $p = 2$, $P = \delta_{11}$. For $p = 5$, P is the product of two invariants*

$$\delta_{11}\delta_{22}\delta_{32}\delta_{41}, \quad \delta_{13}\delta_{24}\delta_{34}\delta_{43}, \tag{50}$$

neither of which is a product of invariants. The last property is true also of the following invariants:

$$\begin{gathered} \gamma_1\delta_{03}, \quad \gamma_4\delta_{02}, \quad \gamma_2\delta_{04}\delta_{12}\delta_{30}\delta_{20}\delta_{42}, \\ \gamma_3\delta_{01}\delta_{10}\delta_{23}\delta_{33}\delta_{40}, \quad \beta\gamma_0\delta_{14}\delta_{21}\delta_{31}\delta_{44}. \end{gathered} \tag{51}$$

The product of these seven invariants and $a\delta_{00}$ equals the product of all the linear functions of a, b, c, d, not proportional modulo 5.

For $p = 2$, each of the 15 linear functions is a factor of just one of the following invariants (no one with an invariant factor):

$$a\delta_{00}, \quad \delta_{11}, \quad \beta\gamma_0\delta_{01}, \quad K = b + c, \quad (a + b + c)\delta_{10}. \tag{52}$$

For any $p \neq 3$, the cubic form has the formal invariant

$$G = 3(bc^p - b^pc) - (ad^p - a^pd), \tag{53}$$

* In those linear factors of the first which lack c, the product of the coefficients of a and b is a quadratic non-residue of 5; in those of the second invariant, a quadratic residue.

and an absolute formal invariant* K of degree $p-1$. For $p=5$,

(54) $K = b^4 + c^4 - b^2d^2 - a^2c^2 - bc^2d - ab^2c + acd^2 + a^2bd.$

Thus, for $p = 5$, K and the discriminant D are invariants of degree 4, and weights $\equiv 0, 2 \pmod 4$, while $a\delta_{00}$ and G are of degree 6 and weight $\equiv 3 \pmod 4$. It follows from § 10 that there are no further invariants of degree less than 8. Now the first and second invariants (51) are of degree 10 and weight $\equiv 1 \pmod 4$. Hence if either is expressible as a polynomial in invariants of lower degrees, it must be the product of D by a linear function of $a\delta_{00}$ and G. This is seen to be impossible either by a consideration of the terms of degree $\geqq 5$ in d or by noting that D has no linear factor. Thus $\gamma_1\delta_{03}$ or $\gamma_4\delta_{02}$ occurs in a fundamental system of invariants.

Invariantive products of linear functions of the coefficients of the cubic form therefore play an important rôle in the theory of its formal invariants. Whether or not they play as dominant a rôle as in the case of the quadratic form is not discussed here. We shall however treat more completely the seminvariants.

10. *Formal Seminvariants of a Binary Cubic for $p > 3$.*—We shall first determine the character of the function to which any seminvariant $S(a, b, c, d)$ reduces when $a = 0$. Set $A = 3b$, $2B = 3c$, $C = d$. Then (26) and (46) give

$$A' = A, \quad B' = A + B, \quad C' = A + 2B + C \qquad (\text{when } a = 0).$$

Any function unaltered by this transformation is (§ 7) a polynomial in A, $B^2 - AC$, γ_0', or the product of such a polynomial by β', where γ_0' and β' are the functions γ_0 and β written in capitals. But

$$\gamma_0' = \prod_{t=0}^{p-1} (3t^2b + 3tc + d) = [\delta_{j0}]_{a=0},$$

$$\beta' = \prod_{t=0}^{p-1} \{\tfrac{3}{2}(2tb + c)\} \equiv [\gamma_k]_{a=0},$$

---

* *Transactions of the American Mathematical Society*, vol. 8 (1907), p. 221; vol. 10 (1909), p. 154, foot-note. *Bulletin of the American Mathematical Society*, vol. 14 (1908), p. 316. Cf. Hurwitz, *l. c.*

modulo p. Hence

$$(55) \qquad S = a\sigma(a, b, c, d) + \gamma_k{}^e\phi(b, q, \delta_{j0}) \qquad (e = 0 \text{ or } 1),$$

where k, j may be given any assigned integral values and

$$(56) \qquad q = c^2 - \tfrac{4}{3}bd, \quad -3b^2q = [D]_{a=0},$$

D being the discriminant of f. We use the seminvariants (II, § 2)

$$(57) \qquad S_2 = -b^2 + ac, \quad S_3 = 2b^3 + a(ad - 3bc).$$

First, let $p = 5$. Then $q \equiv c^2 + 2bd$. We have the formal seminvariants*

$$\sigma_3 = bq - a(ab + 2cd),$$

$$\sigma_4 = K - S_2{}^2 = q^2 + a(abd - 2ac^2 + b^2c + cd^2),$$

$$\sigma_5 = bq^2 + a(-ad^3 - bcd^2 + 3c^3d + abc^2 - 2b^3c + a^3b),$$

$$\sigma_6 = q^3 + a(ad^4 - 2bcd^3 - c^3d^2 + abc^2d - 2b^3cd + a^3bd + 2ac^4$$

$$(58) \qquad - b^2c^3 - 2a^3c^2 + ab^4),$$

$$\sigma_7 = q\gamma_0 + a\{2(b^2 - ac)d^4 + a^2bd^3 - bc^2d^3 - 2c^4d^2 + 2a^2c^2d^2$$

$$- 2ac(b^2 - ac)d^2 - (b^2 - ac)^2d^2 - 2a^4d^2 + 2abc^3d$$

$$+ 2a^3bcd + 2ab^4c + 3(b^2 - ac)c^4 - a^4b^2 + 2a^3c^3\},$$

while $2G$ differs from $b\gamma_0$ by a multiple of a. By (55)–(58), S differs from a polynomial in the seminvariants

$$(59) \quad a, \quad D, \quad S_2, \quad S_3, \quad \sigma_3, \quad K, \quad \sigma_5, \quad \sigma_6, \quad \sigma_7, \quad G, \quad \gamma_0, \quad \delta_{00}$$

by a function $a\lambda + \rho b\delta_{00}^g + \sigma q\delta_{00}^h$, in which ρ and σ are constants at least one of which is zero (in view of the degree of the terms). But the increment to $b\delta_{00}^g$ under transformation (26), (46), is

* As the terms with the factor a were taken all of the proper degree and weight; then a term common to a combination of the seminvariants (59) was deleted. Finally the coefficients were found by a process equivalent to the use of a (non-linear) annihilator, *Transactions of the American Mathematical Society*, vol. 8 (1907), p. 205. Expansions were made in powers of d and the terms involving d rechecked. As each remaining term involves a new coefficient, there is no doubt as to the existence of covariants of type σ_5, σ_6, σ_7, though the terms free of d were not rechecked.

$a\delta_{00}^{g}$ with the term ad^{5g}, while d does not occur to this power in the increment to a function λ of degree $5g$. Again, the increment to $q\delta_{00}^{h}$ has the term $2ad^{1+5h}$, while the increment to a function λ of degree $5h+1$ is of smaller degree in d. Hence $\rho = \sigma = 0$. Then in $a\lambda$, λ is a seminvariant which may be treated as was the initial S.

A fundamental system of formal seminvariants of the binary cubic form modulo 5 *is given by the functions* (59).

11. For $p = 2$, the method of § 10 fails. In place of c we now introduce the seminvariant $K = b + c$. Then the transformation (26), (46), becomes

$$(60)\quad a' = a,\quad K' = K,\quad b' = a + b,\quad d' = a + K + d.$$

By § 3, any seminvariant $S(a, K, b, d)$ becomes for $a = 0$ a polynomial in K, b, $d(K + d)$. In place of the last we may use δ_{00}. Hence

$$S = a\sigma + \phi(b, K, \delta_{00}),\quad \delta_{00} = d(a + K + d).$$

We make use of the seminvariants

$$(61)\quad \begin{gathered} \Delta = ad + bc = \delta_{00} + \delta_{01},\quad \beta = b^2 + ab, \\ \beta + \Delta = bK + a(b + d). \end{gathered}$$

Hence S differs from a polynomial in K, δ_{00}, Δ, β by a function $a\rho + b\tau(\beta, \delta_{00})$. Let (60) replace ρ by ρ'. Then $\rho + \rho' \equiv \tau$ (mod 2). Take $a = K = 0$; then (60) is the identity and $0 \equiv \tau(b^2, d^2)$ identically in b, d. Hence the function $\tau(\beta, \delta)$ is identically zero. Thus $a\rho$ and hence ρ is a seminvariant. Hence a, K, δ_{00}, Δ, β *form a fundamental system of formal seminvariants of the cubic modulo* 2.

Note that Δ^2 is the discriminant, so that Δ is an invariant. The invariants (52) may be expressed in terms of our seminvariants:

$$(62)\quad \begin{gathered} \delta_{11} = I + \Delta,\quad \beta\gamma_0\delta_{01} = \beta(\beta + K^2 + aK)(\Delta + \delta_{00}), \\ (a + K)\delta_{10} = (a + K)(a^2 + I) = a\delta_{00} + KI, \end{gathered}$$

where $I = a^2 + aK + \delta_{00}$ is an invariant.

12. *Miss Sanderson's Theorem.**—Given a modular invariant i of a system of forms under any modular group G, we can construct a formal modular invariant I of the system of forms under G such that $I \equiv i \pmod{p}$ for all integral values of the coefficients of the forms. As the proof does not give a simple method of actually constructing I from i, it is in place here to give a very interesting illustration of the theorem with independent verification. Take as i the fundamental seminvariant $(-1)^m P_{m-1} a_m$ of a binary form f (Lecture II). Then I is the quotient L_{m+1}/L_m, where L_m is given by (16) or (17) with $x_1, \cdots, x_m$ replaced by the first m coefficients $a_0, a_1, \cdots, a_{m-1}$ of the binary form f. Now $x = x' + y'$, $y = y'$, replaces $f(x, y)$ by a form in which the coefficient a_j' is a linear function of $a_0, \cdots, a_j$. Hence L_j is a formal seminvariant of f modulo p. First,

$$\frac{L_2}{L_1} = \begin{vmatrix} a_0^p & a_1^p \\ a_0 & a_1 \end{vmatrix} \div a_0 = a_0^{p-1} a_1 - a_1^p$$

is a formal seminvariant which reduces to $-P_0 a_1$ for integral values of a_0, a_1, where $P_0 = 1 - a_0^{p-1}$. Compare (27). Next,

$$L_3 = \begin{vmatrix} a_0^{p^2} & a_1^{p^2} & a_2^{p^2} \\ a_0^p & a_1^p & a_2^p \\ a_0 & a_1 & a_2 \end{vmatrix},$$

$$C = L_3/L_2 \equiv a_2^{p^2} - a_2^p Q + a_2 L_2^{p-1} \qquad \pmod{p},$$

where, as in (10),

$$Q = \frac{a_0^{p^2} a_1 - a_0 a_1^{p^2}}{L_2} = \sum_{j=0}^{p} a_0^{s(p-j)} a_1^{sj} \quad (s = p - 1).$$

For integral values of the a's, we have

$$L_2 \equiv 0, \quad Q \equiv a_0^s + a_1^s + (p-1) a_0^s a_1^s \equiv 1 - P_1,$$

$$P_1 = (1 - a_0^s)(1 - a_1^s),$$

modulo p, since each term of Q, with $j \neq 0$, $j \neq p$, is congruent

* *Transactions of the American Mathematical Society*, vol. 14 (1913), p. 490.

to $a_0^s a_1^s$. Hence $C \equiv P_1 a_2$. Similarly,

$$L_4/L_3 \equiv -a_3^{p^3} + a_3^{p^2}Q_{32} - a_3^{p}Q_{31} + a_3 L_3^{p-1} \qquad (\text{mod } p),$$

where the Q's are defined by (16) and are congruent to*

$$Q_{31} = Q(L_3/L_2)^{p-1} + L_2^{p^2-p}, \quad Q_{32} = (L_3/L_2)^{p-1} + Q^p,$$

with Q as above. Hence for integral values of the a's,

$$Q_{31} \equiv (1 - P_1)P_1 a_2^{p-1} \equiv 0, \quad Q_{32} \equiv 1 - P_1(1 - a_2^{p-1}) = 1 - P_2,$$

$$L_4/L_3 \equiv -P_2 a_3.$$

13. *Modular Covariants.*—Extending the usual definition of a covariant of an algebraic form f to the case in which the group is the set of all linear transformations with integral coefficients taken modulo p, we obtain the concepts modular covariants or formal modular covariants according as the coefficients of f are integers taken modulo p or are indeterminates. The contrast is the same as in § 5. The universal covariants obtained in § 2 and § 4 do not involve the coefficients of f and hence are formal covariants.

I have recently proved† that *all rational integral modular covariants of any system of modular forms are rational integral functions of a finite number of these covariants.* In the same paper I proved that *a fundamental system of modular covariants of the binary quadratic form* (25) *modulo* 3 *is given by the form f itself, its discriminant Δ, the universal covariants L and Q, together with*‡

$$q = (a + c)(b^2 + ac - 1), \quad f_4 = ax^4 + bx^3y + bxy^3 + cy^4,$$

$$C_1 = (a^2b - b^3)x^2 + 2(b^2 + ac)(c - a)xy + (b^3 - bc^2)y^2, \tag{63}$$

$$C_2 = (\Delta + a^2)x^2 - 2b(a + c)xy + (\Delta + c^2)y^2.$$

Here f_4 is a formal covariant, which is congruent to f for integral

* *Transactions of the American Mathematical Society*, vol. 12 (1911), p. 77.

† *Transactions of the American Mathematical Society*, vol. 14 (1913), pp. 299–310. The extension to cogredient sets of variables has since been made by Professor F. B. Wiley, and will be published in his Chicago dissertation.

‡ No one of the eight is a rational integral function of the remaining seven even in the case of integral coefficients a, b, c taken modulo 3.

values of x, y. Also C_2 and (as here written) C_1 are formal covariants. Note that $-q$ is the invariant (42) of Lecture II. When q is made homogeneous by replacing $-a-c$ by $-a^3-c^3$, we obtain the formal invariant $\Gamma=\gamma_2$, given by (32). The resulting eight formal covariants of f do not form a fundamental system of formal covariants; not all the formal invariants are polynomials in Δ and Γ (§ 8). No instance of a fundamental system of formal covariants has yet been published.

The method of proof will be here illustrated by the new and simpler case of a binary quadratic form (20) with integral coefficients modulo 2. By § 6 any invariant of f is a polynomial in

$$(24') \qquad b, \quad abc, \quad q=(b+1)(a+c)+ac,$$

to which the formal invariants (24) reduce modulo 2. Such a polynomial is congruent to a linear function of these three and unity, since

$$bq \equiv abc \pmod{2}.$$

Further, any seminvariant is a polynomial in a, b and q (§ 6), and hence is a linear function of 1, a, b, ab, q, abc. For,

$$aq \equiv a+ab+abc \pmod{2}.$$

These results are in accord with those obtained otherwise in § 14 of Lecture II. We shall now prove the following theorem:

Every rational integral covariant K of the binary quadratic form f modulo 2 is a rational integral function of f, its invariants b and q, the universal covariants

$$Q=x^2+xy+y^2, \quad L=x^2y+xy^2,$$

and the linear covariant

$$l=(a+b)x+(b+c)y, \quad l^2\equiv f+bQ \pmod{2}.$$

The leading coefficient S of K is a seminvariant and hence is of the form $I+ra+sab$, where r and s are constants, and I is an invariant, a linear combination of the invariants (24′) and unity.

First, let K be of even order $2n$. Then

$$K_1 = K - IQ^n - rf^n - sbf^n$$

is a covariant in which the coefficient of x^{2n} is zero and hence has the factor y. Thus K_1 has the factor L and the quotient is a covariant of order $2n - 3$ to which the next argument applies.

Next, let K be of odd order:

$$K = Sx^{2n+1} + S_1x^{2n}y + \cdots.$$

After subtracting from K constant multiples of lQ^n and blQ^n, in which the coefficients of x^{2n+1} are $a + b$ and $ab + b$, respectively, we may assume that S is an invariant. After also subtracting from K a constant multiple of ILQ^{n-1}, where I is a linear combination of the invariants (24′) and unity, we may assume that $S_1 = \beta_1 a + \beta_2 c$, where the β's are functions of b only. Then the covariance of K with respect to the transformation (21) gives

$$Sx'^{2n+1} + S_1'x'^{2n}y' + \cdots \equiv K \equiv Sx'^{2n+1} + (S + S_1)x'^{2n}y' + \cdots \pmod 2,$$

where S_1' denotes the function S_1 formed for the new coefficients (22). Hence

$$S_1' - S_1 = \beta_2(a + b)$$

must equal the invariant S. Since $\beta_2 b$ is a function of the invariant b, $\beta_2 a$ must be an invariant, so that $\beta_2 = 0$. Thus $S = 0$ and K has the factor L as before. Hence the theorem is true for covariants of order ω if true for those of order $\omega - 3$. But it was proved true for those of order zero.

By a similar method I obtain the following theorem:

A fundamental system of covariants of the binary quadratic form f, given by (20), *and the linear form $\lambda = a_2x + a_1y$ modulo* 2 *is given by f, λ, l,*

$$l_1 = (aa_2 + j)x + (ca_1 + j)y,$$

Q, L and the invariants b, q, $(a_1 - 1)(a_2 - 1)$ and

$$j = (a + b)a_1 + (b + c)a_2.$$

Since a_1 and a_2 are cogredient with x and y, the function j obtained from the covariant l of f is an invariant of f and λ.

The reverse of the last process is important. If we adjoin to a system of binary forms in the variables x' and y' the linear form $yx' - xy'$, any modular invariant of the enlarged system, formal as to x, y, is a modular covariant of the given system with x', y' replaced by x, y. The theorem of § 12 therefore proves the existence of certain formal covariants.*

Applications of Invariants of a Modular Group, §§ 14, 15

14. *Form Problem for the Total Binary Modular Group* Γ.— This group is composed of all binary linear transformations (7) with integral coefficients taken modulo p whose determinant Δ is not divisible by p. By (8),

$$(64)\quad L(x, y) \equiv \Delta L(X, Y), \quad Q(x, y) \equiv Q(X, Y) \pmod{p},$$

so that L^{p-1} and Q are absolute invariants of Γ. Hence, of the functions (11), q is invariant under Γ, while l is unaltered by certain transformations and changed in sign by others. Thus a homogeneous function of q and l having a term which is a power of q is a relative invariant of Γ only when an absolute invariant. Hence if $p > 2$, it involves only even powers of l, and by the homogeneity, only even powers of q. Hence *any absolute invariant of* Γ *is a product of powers of* L^{p-1} *and* Q *by a polynomial in* q^γ, l^γ, *where* $\gamma = 1$ *if* $p = 2$, $\gamma = 2$ *if* $p > 2$.

In particular, L^{p-1} and Q form a fundamental system of absolute invariants of Γ. The so-called form problem for the group Γ requires the determination of all pairs of values of the variables x and y for which L^{p-1} and Q are congruent modulo p to assigned values λ and μ, either integers or imaginary roots of congruences modulo p. We have therefore to solve the system of congruences

$$(65)\qquad \{L(x, y)\}^{p-1} \equiv \lambda, \quad Q(x, y) \equiv \mu \pmod{p}.$$

* After these lectures were delivered, I saw a manuscript by Professor O. E. Glenn, containing tables of formal concomitants for forms of low orders and moduli 2 and 3. He employs transvection between the form and the covariant L of § 2.

First, let $\lambda \not\equiv 0$. For $z \equiv x$ or $z \equiv y$, we have

$$0 \equiv \begin{vmatrix} x^{p^2} & y^{p^2} & z^{p^2} \\ x^p & y^p & z^p \\ x & y & z \end{vmatrix} \equiv Lz^{p^2} - QLz^p + L^p z \qquad (\text{mod } p).$$

Hence x and y are roots of

$$(66) \qquad F(z) = z^{p^2} - \mu z^p + \lambda z \equiv 0 \qquad (\text{mod } p).$$

Having no double root, this congruence has p^2 distinct integral or imaginary roots. These roots are

$$(67) \qquad eX + fY \quad (e, f = 0, 1, \cdots, p-1),$$

where X and Y are particular roots linearly independent modulo p. For,

$$(68) \qquad F(eX + fY) = eF(X) + fF(Y).$$

Hence any pair of solutions x, y of (65) is of the form (7), where $a, \cdots, d$ are integers, whose determinant Δ is not divisible by p, in view of (64_1) and $\lambda \not\equiv 0$.

Conversely, if X and Y are fixed linearly independent solutions of (66), any pair of linear functions of X and Y with integral coefficients, whose determinant is not divisible by p, gives a solution of (65). Indeed, by (68), x and y are solutions of (66). From the two resulting identities, we eliminate λ and μ in turn and get

$$\mu = Q(x, y), \quad \{L(x, y)\}^p = \lambda L(x, y).$$

Since X and Y are linearly independent modulo p, $L(X, Y)$ is not divisible by p [cf. (6)]. Thus $L(x, y) \not\equiv 0$ by (64). Hence (65) hold.

Hence, for $\lambda \not\equiv 0$, the form problem has been reduced to the solution of congruence (66). The latter will be discussed here in the simple but typical* case in which λ and μ are integers. Now the problem to find the real and imaginary roots of a con-

* For the general case, see *Transactions of the American Mathematical Society*, vol. 12 (1911), p. 87.

gruence with integral coefficients is at bottom the problem to factor it into irreducible congruences with integral coefficients.

When v is an integer, $z^p - vz$ is a factor of (66) if and only if v is a root of the characteristic* congruence

$$(69) \qquad v^2 - \mu v + \lambda \equiv 0 \pmod{p}.$$

Such a binomial is a product† of binomials $z^d - \delta$, irreducible modulo p, whose degree d is the exponent to which the integer v belongs modulo p. Since $2p - 1 < p^2$, the function (66) has an irreducible factor $\phi(z)$ of degree $D > 1$, not of the preceding type $z^d - \delta$, and hence with a root r such that r^p/r is not congruent to an integer. Thus every root of (66) is of the form $c_1 r + c_2 r^p$, where the c's are integers. *The irreducible factors of* (66) *are of degree D except those, occurring only when* (69) *has an integral root, of the form $z^d - \delta$, where d is a divisor of D.*

To find D, note that by raising (66) to the powers $p, p^2, \cdots$, we can express z^{p^t} as a linear function l_t of z^p and z. Now D is the least value of t for which $l_t \equiv z$. But the coefficients of l_t are the elements of the first row of the matrix of S^{t-1}, where

$$S = \begin{pmatrix} \mu & -\lambda \\ 1 & 0 \end{pmatrix}.$$

* Note the analogy of (66) with the linear differential equation

$$F(z) = \frac{d^2z}{dt^2} - \mu \frac{dz}{dt} + \lambda z = 0,$$

having the solution $z = e^{vt}$ if v is a root of $v^2 - \mu v + \lambda = 0$. Also, (68) holds. Make dz/dt correspond to z^p and hence d^2z/dt^2 to $(z^p)^p$. Thus the differential equation corresponds to (66), and the integral $z = e^{vt}$ (viz., $dz/dt = vz$) to $z^p = vz$.

† Let $f(z)$ be an irreducible factor of degree d. Its roots are

$$r, \quad r^p \equiv vr, \quad r^{p^2} \equiv v^2 r, \quad \cdots, \quad r^{p^{d-1}} \equiv v^{d-1} r,$$

where $v^d \equiv 1$, $v^l \not\equiv 1$, $0 < l < d$. Thus d is a divisor of $p - 1$. Hence

$$z^{p-1} - v = z^{p-1} - r^{p-1}$$

has the factor $z^d - r^d$. The latter has a root r in common with $f(z)$. But

$$(r^d)^{p-1} \equiv v^d \equiv 1.$$

Thus $\delta = r^d$ is an integer. Hence $f(z) = z^d - \delta$.

But $l_D = z$ implies that $l_{D+1} = z^p$. The condition for the latter is therefore $S^D = 1$. Hence D is the period of S. But (69) is the characteristic determinant of S. According as it has distinct roots v_1 and v_2 or equal roots $v = \frac{1}{2}\mu = \lambda^{\frac{1}{2}}$, a linear substitution of matrix S can be transformed linearly into one of matrix*

$$\begin{pmatrix} v_1 & 0 \\ 0 & v_2 \end{pmatrix}, \quad \begin{pmatrix} v & v \\ 0 & v \end{pmatrix}.$$

According as the characteristic congruence (69) *has distinct* (*real or imaginary*) *roots or a double root, D is the least common multiple of the exponents to which the distinct roots belong modulo p, or is p times the exponent to which the double root belongs.*

Finally, let $\lambda \equiv 0$. By (6), either $y \equiv 0$ or $x - ay \equiv 0 \pmod{p}$, where a is an integer. In the first case,

$$Q = x^{p^2-p}, \quad x^{p^2} - \mu x^p = 0.$$

If $\mu = 0$, then $x = y = 0$. If $\mu \neq 0$, the roots x are equal in sets of p and hence are cx_1 $(c = 0, 1, \cdots, p-1)$, where x_1 is a particular root not divisible by p. In the second case $x - ay \equiv 0$, we take $x - ay$ as a new variable X and conclude from the absolute invariance of Q that

$$Q(x, y) = Q(0, y) = y^{p^2-p}.$$

We thus have the first case with y in place of x.

Using similar methods, I have solved the form problem for the total group of modular linear transformations on m variables.†

15. *Invariantive Classification of Forms.*—Let

$$\phi(x, y) = x^m + \cdots \qquad (m > 1) \tag{70}$$

be a binary form irreducible modulo p and having unity as the coefficient of the highest power of x. Let G be the group of all modular binary linear transformations (1) with integral coef-

* In the second case we use the new variables x and $x - vy$.

† *Transactions of the American Mathematical Society*, vol. 12 (1911), pp. 84–92.

ficients of determinant unity. Let $\phi_1 = \phi, \phi_2, \cdots, \phi_k$ denote all the forms of type (70) which can be transformed into constant multiples of ϕ by transformations of G. Evidently their product $P = \phi_1\phi_2 \cdots \phi_k$ is transformed into c_tP by any transformation t of G. The constant c_t is easily seen* to be congruent to unity. Hence P is an absolute invariant of G. If $m > 2$, no ϕ_i vanishes for a special point. We now apply the theorem in the first part of § 14. Hence, *if $m > 2$, the absolute invariant P is an integral function with integral coefficients of the invariants q, l, each exponent of q and l being even if $p > 2$.* In view of the definition of the ϕ_i, this function of q and l is an irreducible function of those arguments modulo p.

Two binary forms shall be said to be *equivalent* if and only if one of them can be transformed into a constant multiple of the other by a transformation of G. A set of all forms equivalent to a given one shall be called a *genus*. Thus $\phi_1, \cdots, \phi_k$ form a genus. All of the irreducible forms (70) separate into a finite number f of distinct genera; let $P_1, \cdots, P_f$ denote the products of the forms in the respective genera. Thus $\pi_m = P_1 \cdots P_f$ is the product of all of the binary forms $x^m + \cdots$ irreducible modulo p. Hence π_m is a polynomial in q, l with integral coefficients. Hence *the f genera of irreducible binary forms of degree $m > 2$ are characterized invariantively by the f irreducible factors $P_i(q, l)$ of $\pi_m(q, l)$ modulo p.*

We shall see that $\pi_m(q, l)$ is easily computed. By finding its factors irreducible modulo p in the arguments q, l, we shall have invariantive criteria for the equivalence of two irreducible binary forms of degree m. For example, we shall prove that $\pi_3 = q - l$ if $p = 2$, so that all irreducible binary cubic forms modulo 2 are equivalent. Further, $\pi_3 = q^2 - l^2$ if $p > 2$, so that the irreducible cubic factors of $q - l$ are all equivalent, also those of $q + l$, while no factor of the former is equivalent to one of the latter.

* *Transactions of the American Mathematical Society*, vol. 12 (1911), p. 3, § 4. The present section is an account of the simpler topics there treated at length.

In general, let m be a product of powers of the distinct prime numbers $q_1, \cdots, q_\mu$, and set

$$F_t = (x^{p^t}y - xy^{p^t})/L.$$

From the expression for π_m due to Galois we readily obtain

$$\pi_m = \frac{F_m \cdot \Pi F_{m/q_iq_j} \cdot \Pi F_{m/q_iq_jq_kq_l} \cdots}{\Pi F_{m/q_i} \cdot \Pi F_{m/q_iq_jq_k} \cdots},$$

in which the first product in the numerator extends over the $\frac{1}{2}\mu(\mu - 1)$ combinations of $q_1, \cdots, q_\mu$ two at a time, and similarly for the remaining products. By the first theorem of this section, and (11), π_m is a polynomial in

$$J = q^\gamma = Q^{p+1}, \quad K = l^\gamma = L^{p(p-1)} \quad (\gamma = 1 \text{ if } p=2, \gamma=2 \text{ if } p>2).$$

We readily verify the recursion formula

$$F_t \equiv QF_{t-1}^p - KF_{t-2}^{p^2} \qquad (\text{mod } p),$$

since $F_1 = 1$, $F_2 = Q$. In particular,

$$F_3 \equiv J - K, \quad F_4 \equiv Q(F_3{}^p - KJ^{p-1}).$$

Now $\pi_3 = F_3$, $\pi_4 = F_4/Q$. Hence

$$\pi_3 \equiv J - K, \quad \pi_4 \equiv J^p - K^p - KJ^{p-1} \qquad (\text{mod } p).$$

The first of these results was discussed above. Next, for $p = 2$, π_4 is the irreducible quadratic form $q^2 - l^2 - lq$, so that all quartic forms irreducible modulo 2 are equivalent. For $p > 2$, π_4 vanishes for $K = \rho J$, where

$$\rho^p \equiv 1 - \rho \qquad (\text{mod } p).$$

Except for $\rho \equiv \frac{1}{2}$, ρ is a quadratic Galois imaginary since

$$\rho^{p^2} \equiv 1 - \rho^p \equiv \rho \qquad (\text{mod } p).$$

Thus π_4 is a product of $J - 2K$ and $\frac{1}{2}(p - 1)$ irreducible quadratic forms in J, K. Some of the latter yield a quartic in q and l which is irreducible; others yield a quartic which is a product of two irreducible quadratics modulo p. A simple discussion shows that the number of irreducible factors of $\pi_4(q, l)$ is $6k + t + 1$ if $p = 8k + t$ ($t = \pm 1$ or -3), but is $6k + 2$ if $p = 8k + 3$. We have therefore the number f of genera of irreducible quartics modulo p. For quintics and septics, the analogous discussion is simple, for sextics laborious.

We may utilize similarly the invariants (16) of the group on m variables, obtain expressions in terms of them of the product of all forms in m variables of specified types (as quadratic forms transformable into an irreducible binary form, non-vanishing ternary forms, non-degenerate ternary quadratic forms, etc.), and hence draw conclusions as to the equivalence of forms of the specified type.*

* *Transactions of the American Mathematical Society*, vol. 12 (1911), pp. 92–98.

LECTURE IV

MODULAR GEOMETRY AND COVARIANTIVE THEORY OF A QUADRATIC FORM IN m VARIABLES MODULO 2

1. *Introduction.*—The modular form that has been most used in geometry and the theory of functions is the quadratic form

$$q_m(x) = \Sigma c_{ij}x_ix_j + \Sigma b_ix_i^2 \quad (i, j = 1, \cdots, m;\ i < j) \tag{1}$$

with integral coefficients taken modulo 2. In accord with Lecture III, we shall use the term point to denote a set of m ordered elements, not all zero, of the infinite field F_2 composed of the roots of all congruences modulo 2 with integral coefficients. We shall identify such a point $(x_1, \cdots, x_m)$ with $(\rho x_1, \cdots, \rho x_m)$ where ρ is any element not zero in F_2. The point is called real if the ratios of the x's are congruent to integers modulo 2. Let the c_{ij} and b_i in (1) be elements not all zero of the field F_2. Then the aggregate of the points $(x) = (x_1, \cdots, x_m)$ for which $q_m(x) \equiv 0 \pmod 2$ shall be called a quadric locus, in particular, a conic if $m = 3$. The locus is thus composed of an infinitude of points, a finite number of which are real.

While our results are purely arithmetical, we shall find that the employment of the terminology and methods of analytic projective geometry is of great help in the investigation. Usually the proofs are given initially in an essentially arithmetical form. In case a preliminary argument is based upon geometrical intuition, a purely algebraic proof is given later. The geometry brings out naturally the existence of a linear covariant, which is important in the problem of the determination of a fundamental system of covariants.

2. *The Polar Locus.*—The point $(\kappa y_1 + \lambda z_1, \cdots, \kappa y_m + \lambda z_m)$ is on $q(x) \equiv 0$ if

$$\kappa^2 q(y) + \kappa\lambda P(y, z) + \lambda^2 q(z) \equiv 0 \pmod 2, \tag{2}$$

where

$$P(y, z) = \Sigma c_{ij}(y_i z_j + y_j z_i) \quad (i, j = 1, \cdots, m; i < j). \tag{3}$$

If (y) is a fixed point, all points (z) for which $P(y, z) \equiv 0$ (mod 2) are said to form the polar* locus of (y). For $(z) = (y)$, each summand in (3) is congruent to zero modulo 2. Hence the polar of (y) passes through (y). If (z) is on the polar of (y), (2) has a double root $\kappa : \lambda$ and the line joining (y) and (z) is tangent to $q \equiv 0$.

We may write (3) in the form

$$P(y, z) = u_1 y_1 + \cdots + u_m y_m, \tag{3'}$$

where

$$\begin{aligned} u_1 &= \phantom{c_{12}z_1 + c_{23}z_2 + {}} c_{12}z_2 + c_{13}z_3 + c_{14}z_4 + \cdots + c_{1m}z_m, \\ u_2 &= c_{12}z_1 \phantom{{}+ c_{23}z_2} + c_{23}z_3 + c_{24}z_4 + \cdots + c_{2m}z_m, \\ u_3 &= c_{13}z_1 + c_{23}z_2 \phantom{{}+ c_{23}z_3} + c_{34}z_4 + \cdots + c_{3m}z_m, \\ & \cdots\cdots\cdots\cdots\cdots\cdots \\ u_m &= c_{1m}z_1 + c_{2m}z_2 + c_{3m}z_3 + \cdots + c_{m-1m}z_{m-1}. \end{aligned} \tag{4}$$

There is a striking difference between the cases m odd and m even.

3. *Odd Number of Variables; Apex; Linear Tangential Equation.* Let m be odd. Then the determinant of the coefficients in (4) is congruent modulo 2 to a skew-symmetric determinant of odd order and hence is identically congruent to zero. Hence we can find values of $z_1, \cdots, z_m$ not all congruent to zero such that $u_1, \cdots, u_m$ are all zero modulo 2. Thus *the polars of all points* (y) *have at least one point in common.*

We shall limit attention to the case in which the pfaffians

$$C_1 = [23 \cdots m], \quad C_2 = [134 \cdots m], \quad \cdots, \quad C_m = [12 \cdots m-1] \tag{5}$$

are not all congruent to zero. The point $(C_1, \cdots, C_m)$ shall be

* Take $\kappa = 1$ and let (z) be a point not on $q(x) \equiv 0$. Then (2) is a quadratic congruence in λ with coefficients in F_2 and hence has two roots λ_1 and λ_2 in that field. Now the points (y) and (z) are separated harmonically by $(y + \lambda_1 z)$ and $(y + \lambda_2 z)$ if and only if $\lambda_1 \equiv -\lambda_2$, that is, if $\lambda_1 \equiv \lambda_2$ (mod 2). But the condition for a double root of (2) is $P \equiv 0$ (mod 2).

called the *apex** of the locus $q(x) \equiv 0$. Now each $u_i \equiv 0$ if $z_1 \equiv C_1, \cdots, z_m \equiv C_m$. Hence, *for m odd, the polars of all points pass through the apex.*

If (y) is any point not the apex, the line joining (y) to the apex is tangent to $q(x) \equiv 0$ (§ 2). *Thus any line through the apex is tangent to* $q(x) \equiv 0$.

For $m = 3$, it is true conversely that, if the line

$$\Sigma u_i x_i \equiv 0 \pmod 2 \tag{6}$$

is tangent to $q(x) \equiv 0$, it passes through the apex, so that

$$\kappa = \Sigma C_i u_i \tag{7}$$

is zero modulo 2. Taking, for example, $u_3 \not\equiv 0$, we obtain by eliminating x_3 from (6) and $q(x) \equiv 0$ a quadratic equation in x_1 and x_2 whose left member is the square of a linear function modulo 2 if and only if the coefficient of x_1x_2 is congruent to zero. But this coefficient is the product of κ by a power of u_3. *Thus* $\kappa = 0$ *is the tangential equation of* $q(x) \equiv 0$.

The last result is true for any odd m. The spread (6) is said to be tangent to $q(x) \equiv 0$ if the locus of their intersections is degenerate. Taking $u_m \not\equiv 0$, and eliminating x_m between (6) and $q(x) \equiv 0$, we obtain a quadratic form whose discriminant, defined by (24), equals a product of κ by a power of u_m, and hence is degenerate if and only if $\kappa \equiv 0$.

We thus have geometrical evidence that κ *is a formal contravariant of* $q(x)$, i. e., an invariant of $q(x)$ and $\Sigma u_i x_i$.

To give an algebraic proof, note that κ is unaltered when x_i and x_j are interchanged, while

$$x_1 = x_1' + x_2', \quad x_2 = x_2', \quad \cdots, \quad x_m = x_m' \tag{8}$$

replaces $q(x)$ by $q'(x')$ in which the altered coefficients are

$$b_2' = b_2 + b_1 + c_{12}, \quad c_{2i}' = c_{2i} + c_{1i} \quad (i = 3, \cdots, m). \tag{9}$$

* After these lectures were delivered, I learned that Professor U. G. Mitchell had obtained, independently of me, the notion apex ("outside point") for the case $m = 3$, Princeton dissertation, 1910, printed privately, 1913.

The pfaffians $C_2, \cdots, C_m$ are unaltered modulo 2, while

$$(10) \quad C_1' \equiv C_1 + C_2, \quad u_2' \equiv u_2 + u_1, \quad u_i' \equiv u_i \quad (i \neq 2) \qquad (\text{mod } 2).$$

Hence κ is unaltered modulo 2. Note that

$$(11) \qquad \kappa^2 \equiv \begin{vmatrix} 0 & c_{12} & c_{13} & \cdots & c_{1m} & u_1 \\ c_{12} & 0 & c_{23} & \cdots & c_{2m} & u_2 \\ \cdot & \cdot & \cdot & \cdot & \cdot & \cdot \\ c_{1m} & c_{2m} & c_{3m} & \cdots & 0 & u_m \\ u_1 & u_2 & u_3 & \cdots & u_m & 0 \end{vmatrix} \qquad (\text{mod } 2).$$

We saw that $C_1, \cdots, C_m$ are cogredient with $x_1, \cdots, x_m$. This is evident from the fact that the apex is covariantively related to $q(x)$. Hence if we substitute C_1 for $x_1, \cdots, C_m$ for x_m in (1), we obtain the formal invariant

$$(12) \qquad q_m(C) = \Sigma c_{ij} C_i C_j + \Sigma b_i C_i^2 \quad (i, j = 1, \cdots, m; i < j).$$

If this invariant vanishes, the apex is on the locus, which is then a cone. Indeed, by (2), every point on the line joining (C) to a point on $q(x) \equiv 0$ lies on the latter. Hence $q(x)$ can be transformed into a form in $m - 1$ variables and hence has the discriminant zero. To argue algebraically, let new variables be chosen so that the apex becomes $(0, \cdots, 0, 1)$. The polar of any point (y) passes through the apex. Taking $z_1 = 0, \cdots, z_{m-1} = 0$, $z_m = 1$ in (4), we see that the polar (3′) becomes $c_{1m}y_1 + \cdots + c_{m-1m}y_{m-1}$, which must vanish for arbitrary y's. Hence $b_m x_m^2$ is the only term of (1) involving x_m. But the apex is on the locus. Hence $b_m = 0$ and $q(x)$ is free of x_m. The converse is obvious from (5).

Whether m is odd or even, $q(x)$ has the invariant

$$(13) \qquad A_m = \Pi(c_{ij} + 1) \quad (i, j = 1, \cdots, m; i < j).$$

This is evidently true by (9) or as follows. If $A_m \equiv 1 \pmod 2$, every $c_{ij} \equiv 0$ and $q \equiv (\Sigma b_i x_i)^2$; while if $A_m \equiv 0$, at least one c_{ij} is not congruent to zero, and q is not a double line.

Hence the product $A_m q(x)$ is a covariant; in fact, the square of the linear covariant $A_m \Sigma b_i x_i$. We shall see however that there exists a more fundamental linear covariant.

4. *Covariant Line of a Conic.*—Since we shall later treat in detail the case $m = 3$, we shall replace (1) by the simpler notation

$$(14) \quad F(x) = a_1 x_2 x_3 + a_2 x_1 x_3 + a_3 x_1 x_2 + b_1 x_1^2 + b_2 x_2^2 + b_3 x_3^2.$$

Its apex is (a_1, a_2, a_3). Its discriminant (12) is

$$(15) \quad \Delta = F(a_1, a_2, a_3) \equiv a_1 a_2 a_3 + a_1^2 b_1 + a_2^2 b_2 + a_3^2 b_3.$$

The invariant (13) becomes

$$(16) \qquad A = \alpha_1 \alpha_2 \alpha_3 \qquad (\alpha_i = a_i + 1).$$

Consider a form (14) with integral coefficients and not the square of a linear function. Then not every a_i is congruent to zero modulo 2. By an interchange of variables we may set $a_3 \equiv 1$. Replace x_1 by $X_1 + a_1 x_3$ and x_2 by $X_2 + a_2 x_3$. We get

$$X_1 X_2 + b_1 X_1^2 + b_2 X_2^2 + \Delta x_3^2.$$

Let $\Delta \equiv 1$. Replace x_3 by $X_3 + b_1 X_1 + b_2 X_2$. We get

$$(17) \qquad \phi = X_1 X_2 + X_3^2.$$

The only real points on $\phi \equiv 0 \pmod 2$ are (1, 1, 1), (1, 0, 0), (0, 1, 0). In addition to these and the apex (0, 0, 1), the only real points in the plane are (1, 1, 0), (0, 1, 1), (1, 0, 1). These lie on the straight line

$$(18) \qquad X_1 + X_2 + X_3 \equiv 0 \qquad \pmod 2.$$

Hence with every non-degenerate conic modulo 2 is associated covariantly a straight line.

The inverse of the transformation used above is

$$X_1 = x_1 + a_1 x_3, \quad X_2 = x_2 + a_2 x_3,$$

$$X_3 = b_1 x_1 + b_2 x_2 + (1 + a_1 b_1 + a_2 b_2) x_3.$$

It must therefore replace ϕ by the general form (14) having

$a_3 \equiv \Delta \equiv 1$. It actually replaces (18) by

$$(b_1 + 1)x_1 + (b_2 + 1)x_2 + (b_3 + \alpha_1\alpha_2 + 1)x_3,$$

in which we have added $\Delta + 1 \equiv 0$ to the initial coefficient of x_3. Guided by symmetry, we restore terms which become zero for $a_3 = 1$ and get

$$(19) \quad L = \sum_{i=1}^{3} (\beta_i + 1)x_i, \qquad \beta_1 = b_1 + \alpha_2\alpha_3, \quad \beta_2 = b_2 + \alpha_1\alpha_3, \quad \beta_3 = b_3 + \alpha_1\alpha_2.$$

Making the terms homogeneous we obtain the formal covariant

$$(20) \qquad L = B_1x_1 + B_2x_2 + B_3x_3,$$

$$(21) \quad B_1 = b_1^2 + a_2a_3 + a_2^2 + a_3^2, \quad B_2 = b_2^2 + a_1a_3 + a_1^2 + a_3^2, \quad B_3 = b_3^2 + a_1a_2 + a_1^2 + a_2^2.$$

Under the substitution $(a_ia_j)(b_ib_j)$ induced upon the coefficients of F by (x_ix_j), we see that B_i and B_j are interchanged. Under (9), viz.,

$$(22) \qquad b_2' \equiv b_2 + b_1 + a_3, \quad a_1' \equiv a_1 + a_2 \pmod 2,$$

there results

$$(23) \qquad B_1' \equiv B_1, \quad B_2' \equiv B_2 + B_1, \quad B_3' \equiv B_3 \pmod 2.$$

Hence (20) *is a formal covariant of* F. For other interpretations of L see § 8.

5. *Even Number of Variables.*—The determinant of the coefficients in (4) is congruent modulo 2 to the square of the pfaffian

$$(24) \qquad \Delta_m = [123 \cdots m].$$

This is in fact the discriminant of q_m, which is degenerate if and only if $\Delta_m \equiv 0 \pmod 2$. I have elsewhere* discussed at length the invariants of q_m.

* *Transactions of the American Mathematical Society*, vol. 8 (1907), p. 213 (case $m = 2$); vol. 10 (1909), pp. 133–149; *American Journal of Mathematics*, vol. 30 (1908), p. 263; *Proccedings of the London Mathematical Society*, (2), vol. 5 (1907), p. 301.

If $\Delta_m \not\equiv 0 \pmod 2$, we can solve equations (4) for the z's. Substituting the resulting values into $q(z)$, we obtain the tangential equation $U_m \equiv 0$ of $q(x) \equiv 0$. For $m = 2$ and $m = 4$, we get

$$(25)\qquad \begin{aligned} U_2 &= c_{12}u_1u_2 + b_2u_1^2 + b_1u_2^2, \\ U_4 &= [1234]\Sigma c_{34}u_1u_2 + \Sigma(c_{23}c_{24}c_{34} + b_2c_{34}^2 + b_3c_{24}^2 + b_4c_{23}^2)u_1^2. \end{aligned}$$

Bordering the algebraic discriminant of (1), we find that

$$(26)\qquad 2U_m \equiv \begin{vmatrix} 2b_1 & c_{12} & c_{13} & \cdots & c_{1m} & u_1 \\ c_{12} & 2b_2 & c_{23} & \cdots & c_{2m} & u_2 \\ \cdot & \cdot & \cdot & \cdot & \cdot & \cdot \\ c_{1m} & c_{2m} & c_{3m} & \cdots & 2b_m & u_m \\ u_1 & u_2 & u_3 & \cdots & u_m & 0 \end{vmatrix} \pmod 4.$$

Finally, let $\Delta_m \equiv 0 \pmod 2$. Then all of the first minors of the matrix of the coefficients in (4) are zero modulo 2. Hence the polars of all points have in common the points of a straight line S. Since its discriminant vanishes, $q(x)$ can be transformed linearly into a quadratic form in $x_1, \cdots, x_{m-1}$, which therefore represents a cone with the vertex $(0, \cdots, 0, 1)$. Let (z) be the vertex of the initial cone $q(x) \equiv 0$. If (x) is any point on the cone, $(x + \lambda z)$ is on the cone, and, by (2), $P(x, z)$ is congruent to zero identically in $x_1, \cdots, x_m$. Hence the linear functions (4) all vanish. Thus the line S meets the cone in its vertex, and z_m^2 is the discriminant of $q_{m-1}(x)$, while z_i^2 is obtained from that discriminant by interchanging m and i. For example, if $m=4$,

$$z_4^2 = c_{12}c_{13}c_{23} + b_1c_{23}^2 + b_2c_{13}^2 + b_3c_{12}^2, \cdots,$$

$$z_1^2 = c_{23}c_{24}c_{34} + b_2c_{34}^2 + b_3c_{24}^2 + b_4c_{23}^2.$$

The product of the general form (1) by $\delta \equiv \Delta_m + 1$ is a quadratic form whose discriminant is zero modulo 2 and hence has the vertex $(\delta z_1, \cdots, \delta z_m)$, where z_i^2 has the value just given. Hence $\delta z_1^2, \cdots, \delta z_m^2$ are cogredient with $x_1, \cdots, x_m$.

6. *Covariant Plane of a Degenerate Quadric Surface.*—The product of q_4 by $\delta = [1234] + 1$ is a quaternary form f whose

discriminant is zero and hence can be transformed into a form (14) free of x_4. With this cone $F \equiv 0$ is associated covariantively the plane $l = 0$, where l is the ternary covariant (19). Hence f has a linear covariant L which reduces to l when $b_4 \equiv 0$, $c_{i4} \equiv 0$ $(i = 1, 2, 3)$. Relying upon symmetry and the presence of the factor δ, we are led to conjecture that

$$(27)\quad \begin{aligned} L = \delta\{b_1 + 1 + (c_{12} + 1)(c_{13} + 1)(c_{14} + 1)\}x_1 + \cdots \\ + \delta\{b_4 + 1 + (c_{14} + 1)(c_{24} + 1)(c_{34} + 1)\}x_4. \end{aligned}$$

It is readily verified algebraically that L is a covariant of q_4.

There is a simple interpretation of L. If $[1234] \not\equiv 0 \pmod 2$, then $\delta \equiv 0$ and L is identically zero. If $[1234] \equiv 0$, q_4 is degenerate and can be transformed into $\phi = x_1x_2 + x_3^2$ or a form involving only x_1 and x_2. In the former case, $L = x_1 + x_2 + x_3$. Of the 15 real points in space, the seven $(100x)$, $(010x)$, $(111x)$ and (0001) are on the cone $\phi \equiv 0$, the two $(001x)$ are on the invariant line S through the vertex (0001) of the cone and the apex (0010) of the conic cut out by $x_4 \equiv 0$, while the remaining six $(101x)$, $(011x)$, $(110x)$ lie on the plane $L = 0$. Hence with a degenerate quadric surface, not a pair of planes, is associated covariantively a plane, just as a line (19) is associated with a non-degenerate conic (14).

Every linear covariant is of the form IL, where I is an invariant. Every quadratic covariant is a linear combination of the IL^2 and Iq_4.

7. *A Configuration Defined by the Quinary Surface.*—A q_5 whose discriminant is not zero modulo 2 can be transformed into

$$F = x_1x_2 + x_3x_4 + x_5^2.$$

The 15 real points on $F \equiv 0 \pmod 2$ are given in the last column of the table below. In addition to these and the apex (00001) of F, there are just 15 real points in space:

$1=(00011)$, $2=(01001)$, $3=(01011)$, $4=(00101)$, $5=(01101)$,
$6=(00110)$, $7=(01110)$, $8=(10001)$, $9=(10011)$, $a=(10101)$,
$b=(10110)$, $c=(11000)$, $d=(11010)$, $e=(11100)$, $f=(11111)$.

These lie by threes in exactly 20 straight lines, which occur in the columns of the table, with the heading " Sides." With these lines we can form exactly 15 complete quadrilaterals, the three diagonals of each of which intersect* in a point on $F \equiv 0$, given in the last column. The columns, with the heading "Plane," give the equations defining the plane of the quadrilateral. In each case, the two equations of the plane have in common with $F \equiv 0$ a single real point, the intersection of the diagonals. Thus the real points on $F \equiv 0$ are its points of contact with these tangent planes.

Sides				Diagonals			Plane	Inter-section
146	157	356	347	13	45	67	$x_1=0,\quad x_3+x_4+x_5=0$	01000
146	1*ab*	49*b*	69*a*	19	4*a*	6*b*	$x_2=0,\quad x_3+x_4+x_5=0$	10000
146	1*ef*	4*df*	6*de*	1*d*	4*e*	6*f*	$x_1=x_2=x_3+x_4+x_5$	11001
157	1*ab*	5*ac*	7*bc*	1*c*	5*b*	7*a*	$x_1+x_2+x_3=x_3+x_4+x_5=0$	11011
157	1*ef*	58*e*	78*f*	18	5*f*	7*e*	$x_2=x_3,\quad x_1=x_3+x_4+x_5$	10010
1*ab*	1*ef*	2*ae*	2*bf*	12	*af*	*be*	$x_1=x_3,\quad x_2=x_3+x_4+x_5$	01010
28*c*	29*d*	38*d*	39*c*	23	89	*cd*	$x_3=0,\quad x_1=x_2+x_5$	00010
28*c*	2*ae*	5*ac*	58*e*	25	8*a*	*ce*	$x_4=0,\quad x_1=x_2+x_5$	00100
28*c*	2*bf*	78*f*	7*bc*	27	8*b*	*cf*	$x_3=x_4,\quad x_1=x_2+x_4+x_5$	00111
29*d*	2*ae*	69*a*	6*de*	26	9*e*	*ad*	$x_1=x_3+x_4=x_2+x_5$	01111
29*d*	2*bf*	49*b*	4*df*	24	9*f*	*bd*	$x_1=x_4,\quad x_2=x_3+x_4+x_5$	01100
347	38*d*	4*df*	78*f*	3*f*	48	7*d*	$x_2=x_4,\quad x_1=x_3+x_4+x_5$	10100
347	39*c*	49*b*	7*bc*	3*b*	4*c*	79	$x_4=x_1+x_2=x_3+x_5$	11101
356	38*d*	58*e*	6*de*	3*e*	5*d*	68	$x_2=x_3+x_4,\quad x_1=x_2+x_5$	10111
356	39*c*	5*ac*	69*a*	3*a*	59	6*c*	$x_5=x_1+x_2=x_3+x_4$	11110

8. *Certain Formal and Modular Covariants of a Conic.*—For conic (14), the polar form is

$$
\begin{vmatrix} a_1 & a_2 & a_3 \\ y_1 & y_2 & y_3 \\ z_1 & z_2 & z_3 \end{vmatrix} . \tag{28}
$$

Hence if two sets of variables y_i and z_i be transformed cogrediently with the set x_i, this polar form (28) is a covariant of F and the two points (y), (z), in an extended sense of the term

* The dual of the theorem of Veblen and Bussey, " Finite projective geometries," *Transactions of the American Mathematical Society*, vol. 7 (1906), p. 245.

covariant. In particular, if we take $(y) = (x)$, $(z) = (x^{2^n})$, we obtain a covariant of F in the narrow sense used in these lectures. In particular,

$$(29) \qquad K = \begin{vmatrix} a_1 & a_2 & a_3 \\ x_1 & x_2 & x_3 \\ x_1^2 & x_2^2 & x_3^2 \end{vmatrix}, \qquad M = \begin{vmatrix} a_1 & a_2 & a_3 \\ x_1 & x_2 & x_3 \\ x_1^4 & x_2^4 & x_3^4 \end{vmatrix}$$

are formal covariants of F. While the discriminant Δ, given by (15), is a formal invariant, (16) is not. But

$$(30) \qquad A + \Delta + 1 \equiv \alpha \pmod{2},$$

$$(31) \qquad \alpha = \Sigma a_i b_i + \Sigma a_i^2 + a_1 a_2 + a_1 a_3 + a_2 a_3,$$

α being a formal invariant of F. By (23), the B's are contragredient to the x's and hence to the a's, so that

$$(32) \qquad \Delta_1 = \Sigma a_i B_i = \Sigma a_i b_i^2 + \Sigma a_i a_j^2 + a_1 a_2 a_3$$

is a formal invariant. For integral values of a_i, b_i,

$$(33) \qquad \Delta_1 \equiv \Delta \equiv \Sigma a_i(\beta_i + 1) \pmod{2}.$$

Any form with undetermined integral coefficients $c_1, c_2, \cdots$, taken modulo 2, has, by (21) of Lecture I, the invariant $(c_1 + 1)(c_2 + 1) \cdots$. Thus (16) is an invariant of (7) and hence of F. Likewise from (19) and F itself, we obtain the invariants

$$(34) \qquad J = \beta_1 \beta_2 \beta_3, \quad AJ = A\Pi(b_i + 1).$$

In (6) we made use geometrically of

$$(35) \qquad \lambda = u_1 x_1 + u_2 x_2 + u_3 x_3.$$

Now $F + t\lambda^2$ is congruent modulo 2 to the quadratic form derived from F by replacing each b_i by $b_i + tu_i^2$. Making this replacement in Δ, we see that the coefficient of t is congruent to κ^2, where

$$(36) \qquad \kappa = a_1 u_1 + a_2 u_2 + a_3 u_3$$

is therefore a formal invariant* of F and λ. Making the same

* Since (36) is a contravariant of F, $\Sigma a_i(\partial C/\partial x_i)$ is a covariant of F if C is. Taking Q_2, Q_1, L as C, we get K, M, Δ, respectively.

replacement in J and taking t and u_i to be integers, we obtain as the coefficient of $t \equiv t^2$

$$(37) \quad w = \beta_1\beta_2u_3 + \beta_1\beta_3u_2 + \beta_2\beta_3u_1 + \beta_1u_2u_3 + \beta_2u_1u_3 + \beta_3u_1u_2 + u_1u_2u_3,$$

a modular invariant of F and λ. By the theorem used above,

$$(38) \quad u = (u_1 + 1)(u_2 + 1)(u_3 + 1)$$

is an invariant of λ. In $w + u + 1$, we replace β_i by the congruent value $B_i + 1$, and render the expression homogeneous in the u's and B's separately. We get

$$(39) \quad \omega = \Sigma(B_1B_2 + B_1^2 + B_2^2)u_3^2 + \Sigma B_1^2u_2u_3,$$

a formal invariant of F, λ. For, it is unaltered by the substitution

$$(a_ia_j)(b_ib_j)(u_iu_j),$$

induced by (x_ix_j), and by the substitution (23) and (10) induced by (8). Let the coefficients of F be integers not all even. Then (39) becomes

$$(39') \quad \Sigma(\beta_1\beta_2 + 1)u_3^2 + \Sigma(\beta_1 + 1)u_2u_3.$$

Its covariant L is identically zero. Hence, by the table in § 9, if ω is not identically zero it can be transformed into $u_1^2 + u_2^2 + u_1u_2$ and hence vanishes for a single set of integral values of u_1, u_2, u_3. These are seen to be $u_i = \beta_i + 1$. Hence* *the line* $L = 0$ *is the only line with integral line coordinates on the line locus* (39).

The invariant A for (39) is J (its discriminant is zero, as just seen). Thus a knowledge of any one of the concomitants L, J, ω implies that of the other two.

The covariance of K in (29) implies that

$$(40) \quad \xi_1 = \begin{vmatrix} x_2 & x_3 \\ x_2^2 & x_3^2 \end{vmatrix}, \qquad \xi_2 = \begin{vmatrix} x_1 & x_3 \\ x_1^2 & x_3^2 \end{vmatrix}, \qquad \xi_3 = \begin{vmatrix} x_1 & x_2 \\ x_1^2 & x_2^2 \end{vmatrix}$$

---

* Also thus: just as the point conic $F \equiv 0$ determines its line equation (36) and hence its apex (a), so the covariant line conic (39) determines the point equation $\Sigma B_i^2x_i \equiv 0$, which is the line $L \equiv 0$ for integral values of the coefficients.

are contragredient with a_1, a_2, a_3 and hence with x_1, x_2, x_3, and therefore cogredient with u_1, u_2, u_3. Thus (39) yields the formal covariant

$$(41') \qquad W' = \Sigma(B_1B_2 + B_1^2 + B_2^2)\xi_3^2 + \Sigma B_1^2\xi_2\xi_3.$$

From this or (39′), we obtain the modular covariant

$$(41) \qquad W = \Sigma(\beta_1\beta_2 + 1)\xi_3^2 + \Sigma(\beta_1 + 1)\xi_2\xi_3.$$

In these notations (29) become

$$(42) \qquad K = \Sigma a_i\xi_i, \quad M = \Sigma a_1\xi_1(x_2^2 + x_2x_3 + x_3^2).$$

Finally, by (16) of Lecture III, we have the universal covariants

$$(43) \quad L_3 = \begin{vmatrix} x_1 & x_2 & x_3 \\ x_1^2 & x_2^2 & x_3^2 \\ x_1^4 & x_2^4 & x_3^4 \end{vmatrix}, \quad \begin{aligned} Q_1 &= \Sigma x_1^4x_2^2 + \Sigma x_1^4x_2x_3 + x_1^2x_2^2x_3^2, \\ Q_2 &= \Sigma x_1^4 + \Sigma x_1^2x_2^2 + x_1x_2x_3\Sigma x_1. \end{aligned}$$

The covariant line $L \equiv 0$ of a non-degenerate conic $F \equiv 0$ is determined by the three (collinear) diagonal points of the complete quadrangle having as its vertices the apex (a) and the three intersections of $F \equiv 0$ with its covariant cubic curve $K \equiv 0$.

Fundamental System of Covariants of the Ternary Form F, §§ 9–32

9. *Invariants of F.*—A fundamental system of invariants of F is given by Δ, A, J. It suffices to prove that they completely characterize the classes of forms F under the group of all ternary linear transformations with integral coefficients modulo 2. This is evident from the following table

Class	Δ	A	J	L
$x_1x_2 + x_3^2$	1	0	0	$x_1 + x_2 + x_3$
$x_1x_2 + x_1^2 + x_2^2$	0	0	1	0
x_1x_2	0	0	0	$x_1 + x_2$
x_1^2	0	1	0	x_1
0	0	1	1	0

As to the classes, we saw in § 4 that, if F is not the square of a linear function (i. e., not reducible to x_1^2 or 0), it can be transformed into $x_1x_2 + b_1x_1^2 + b_2x_2^2 + \Delta x_3^2$ and hence into one of the first three classes of the table. By means of the relations

$$(44)\quad \Delta A \equiv 0, \quad \Delta J \equiv 0, \quad \Delta^2 \equiv \Delta, \quad A^2 \equiv A, \quad J^2 \equiv J \qquad (\text{mod } 2),$$

any polynomial in Δ, A, J equals a linear function of

$$(45)\qquad 1, \quad \Delta, \quad A, \quad J, \quad AJ.$$

These are linearly independent since there are five classes.

10. *Leader of a Covariant of F.*—Let S be the coefficient of x_3^ω in a covariant of order ω of F. Writing (14) in the form

$$(46)\quad F = f + lx_3 + b_3x_3^2, \quad f = b_1x_1^2 + a_3x_1x_2 + b_2x_2^2, \quad l = a_2x_1 + a_1x_2,$$

we see that the leader S is a function of b_3 and the invariants of the pair of forms f and l under the linear group on x_1, x_2.

In the modular covariants forming a fundamental system for f (§ 13 of Lecture III), we replace x_1 by a_1 and x_2 by a_2 and obtain a fundamental system of modular invariants of the pair f and l:

$$(47)\quad a_3, \quad \alpha_1\alpha_2, \quad q = b_1b_2 + (b_1 + b_2)\alpha_3, \quad j = (b_1 + a_3)a_1 + (b_2 + a_3)a_2,$$

where $\alpha_i = a_i + 1$. By means of the relations

$$(48)\qquad \alpha_1\alpha_2 j \equiv 0, \quad qj \equiv j + a_3 j \qquad (\text{mod } 2),$$

any polynomial in the four functions (47) can be reduced to a linear combination of

$$(49)\quad 1, \quad a_3, \quad q, \quad a_3q, \quad \alpha_1\alpha_2, \quad \alpha_1\alpha_2a_3, \quad \alpha_1\alpha_2q, \quad \alpha_1\alpha_2a_3q, \quad j, \quad a_3j.$$

These form a complete set of linearly independent* invariants of f, l.

* Instead of verifying as usual that these 10 functions are linearly independent, we may deduce that result from the fact that there are 10 classes:

$$l = x_1, \quad f = a_3x_1x_2 + \alpha_3x_2^2 \quad \text{or} \quad qx_1^2 + a_3x_1x_2 + a_3x_2^2,$$

$$l \equiv 0, \quad f = x_1^2 + x_1x_2 + x_2^2, \quad x_1x_2, \quad x_1^2 \quad \text{or} \quad 0.$$

Since (47) characterize the classes, they form a fundamental system.

Hence S is a linear combination of the functions (49) and their products by b_3. Moreover, S must remain unaltered modulo 2 when a_3 and b_1 are replaced by

$$(50) \qquad a_3' \equiv a_3 + a_1, \quad b_1' \equiv b_1 + b_3 + a_2,$$

which are the only altered coefficients of the form obtained from F by the transformation

$$(51) \qquad x_1 \equiv x_1', \quad x_2 \equiv x_2', \quad x_3 \equiv x_3' + x_1' \qquad (\text{mod } 2).$$

Both requirements are evidently met by the functions

$$(52) \qquad 1, \quad \alpha_1\alpha_2, \quad b_3, \quad b_3\alpha_1\alpha_2$$

and any invariant of F. We find that

$$A = \alpha_1\alpha_2(a_3 + 1), \quad \Delta = \alpha_1\alpha_2 a_3 + j + a_3 b_3 + a_3,$$

$$(53) \qquad J = \alpha_1\alpha_2(a_3 + 1)(b_3 + 1) + b_3 j + a_3 b_3 j + b_3 q + \alpha_1\alpha_2 q,$$

$$AJ = \alpha_1\alpha_2(a_3 + 1)(b_3 + 1)(q + 1).$$

From these and their products by b_3, we see that

$$(54) \qquad AJ, \quad b_3 J, \quad J, \quad b_3\Delta, \quad b_3 A, \quad \Delta, \quad A$$

contain the respective terms

$$b_3\alpha_1\alpha_2 a_3 q, \quad b_3\alpha_1\alpha_2 q, \quad \alpha_1\alpha_2 q, \quad b_3 j, \quad b_3\alpha_1\alpha_2 a_3, \quad j, \quad \alpha_1\alpha_2 a_3,$$

while no one involves an earlier one of these terms. Hence any linear combination of the functions (49) and their products by b_3 is a linear combination of the functions (52), (54) and

$$(55) \quad a_3, \quad b_3 a_3, \quad q, \quad b_3 q, \quad a_3 q, \quad b_3 a_3 q, \quad a_3 j, \quad b_3 a_3 j, \quad \alpha_1\alpha_2 a_3 q.$$

A linear combination of the latter is of the form

$$\sigma = m_1 a_3 + m_2 q + m_3 a_3 q + m_4 a_3 j + m\alpha_1\alpha_2 a_3 q,$$

where $m_1, \cdots, m_4$ are linear functions of b_3, while m is a constant. The coefficient of $a_3 b_1$ is seen to be

$$p = m_2 + m_3 b_2 + m_4 a_1 + m b_2 \alpha_1 (R + b_3 + 1),$$

where $R \equiv b_3 + a_2$ is the increment to b_1 in (50). Set

(56) $\sigma = pa_3b_1 + ra_3 + sb_1 + t$ ($p, \cdots, t$ independent of a_3, b_1).

Let the substitution (50) replace σ by σ'. Then

(57) $$\sigma' - \sigma = pRa_3 + pa_1b_1 + pa_1R + ra_1 + sR.$$

This is zero for every a_3, b_1 if and only if

(58) $$pR \equiv 0, \quad pa_1 \equiv 0, \quad ra_1 \equiv sR \qquad (\text{mod } 2).$$

For $p = m_2 + \cdots$, $pa_1 \equiv 0$ gives $m_3 \equiv 0$, $m_4 \equiv m_2$. Then $pR \equiv 0$ gives $m_2 \equiv 0$, $mb_3 \equiv 0$, whence $m \equiv 0$. Thus $\sigma = m_1a_3$, so that $m_1 \equiv 0$. Hence *the leader of a covariant of F has the form*

(59) $$I + b_3I_1 + c\alpha_1\alpha_2 + d\alpha_1\alpha_2b_3,$$

where I and I_1 are invariants, c and d are constants.

Covariants Whose Leaders are Not Zero, §§ 11–19

11. Consider a covariant of odd order ω:

(60) $$C = Sx_3^{\omega} + S_1x_3^{\omega-1}x_1 + S_2x_3^{\omega-2}x_1^2 + \cdots.$$

If S_1' is derived from S_1 by the substitution (50), then, by (51),

(61) $$S_1' \equiv S_1 + \omega S \equiv S_1 + S \qquad (\text{mod } 2).$$

Give S_1 the notation (56). Then S is given by (57) and has no term with the factor a_3b_1. Now a_3b_1 enters no term of (59) except J and AJ of I and* b_3J of b_3I_1, and in these is multiplied by

(62) $$b_3\alpha_1 + \alpha_1\alpha_2, \quad \alpha_1\alpha_2(b_2 + 1)(b_3 + 1), \quad b_3\alpha_1a_2,$$

respectively. Since the latter are linearly independent, neither J nor AJ occurs in the I, I_1 of the leader (59). Also, A and $\alpha_1\alpha_2$ occur only in the combinations $A + 1$, $\alpha_1\alpha_2 + 1$, since (57) has no constant term. The coefficients of x_3^{ω} in L^{ω}, AL^{ω}, $(A + \Delta)L^{\omega}$ are respectively

(63) $$b_3 + \alpha_1\alpha_2 + 1, \quad Ab_3, \quad \Delta + \Delta b_3 + b_3\alpha_1\alpha_2,$$

* AJ is not retained in I_1, since $b_3AJ \equiv 0$, AJ being (34).

where L is the linear covariant (19). After subtracting from C a linear combination of these three covariants, we may set

$$S = m_1(A + 1) + m_2\Delta + m_3b_3 + mb_3\alpha_1\alpha_2.$$

Since $\beta_3b_3\alpha_1\alpha_2 \equiv 0$, $\Delta J \equiv 0$, the leader of the covariant JC is

$$JS = m_1AJ + m_1J + m_3b_3J.$$

Hence $m_1 \equiv m_3 \equiv 0$. The coefficient of a_3 in S is now $m_2(a_1a_2 + b_3)$ and must vanish for $b_3 \equiv a_2$ since it is of the form pR by (57). Hence $m_2 \equiv 0$. Thus $S = mb_3\alpha_1\alpha_2$. For $\omega > 1$, $mFL^{\omega-2}$ has this same leader. For $\omega = 1$,

$$C = m(b_3\alpha_1\alpha_2x_3 + b_1\alpha_2\alpha_3x_1 + b_2\alpha_1\alpha_3x_2),$$

which satisfies (61) only when $m = 0$. Hence *every linear covariant is a linear function of L, AL, ΔL; every covariant of odd order $\omega > 1$ differs from a linear combination of L^ω, AL^ω, ΔL^ω, $FL^{\omega-2}$ by a covariant whose leader is zero.*

12. In the covariants of order $4n$

(64) $$IQ_2^n,\quad IF^{2n},\quad L^{4n},\quad F^{2n-1}L^2 \quad (I \text{ an invariant}),$$

the coefficients of x_3^{4n} are respectively

$$I,\quad b_3I,\quad b_3 + \alpha_1\alpha_2 + 1,\quad b_3\alpha_1\alpha_2.$$

Linear combinations of these give every leader (59). Hence *every covariant of order $4n$ differs from a linear combination of the covariants* (64) *by a covariant whose leader is zero.*

13. In the covariants of order $\omega = 4n + 2$

(65) $$IQ_2^nF,\quad Q_2^nL^2,\quad \Delta Q_2^nL^2 \qquad (I \text{ an invariant}),$$

the coefficients of x_3^ω are respectively

$$b_3I,\quad b_3 + \alpha_1\alpha_2 + 1,\quad \Delta + b_3(\Delta + \alpha_1\alpha_2a_3).$$

The sum of the third function and $b_3(A + \Delta)$ is $\Delta + b_3\alpha_1\alpha_2$. Hence any covariant C is of the form $P + C'$, where P is a linear combination of the covariants (65), while C' is a covariant whose leader is an invariant. For $\omega = 2$,

$$C' = Sx_3^2 + S_1x_3x_1 + Sx_1^2 + x_2\phi.$$

This is transformed by (51) into a function having S_1 as the coefficient of $x_1'^2$. Since S is an invariant, $S_1 = S$. Thus every coefficient of C' equals S. Then (51) transforms C' into a function in which the coefficient of $x_1'x_2'$ is zero, so that $S = 0$. Hence *every quadratic covariant is a linear function of*

$$(66) \qquad F, \quad AF, \quad \Delta F, \quad JF, \quad L^2, \quad \Delta L^2.$$

14. There remains the more difficult case of covariants (60) of order $\omega = 4n + 2 > 2$. If S_i' is the function obtained from S_i by the substitution (50), then

$$(67) \qquad S_1' = S_1, \quad S_2' = S + S_1 + S_2.$$

Now S_1 is unaltered also by the substitutions (22) and

$$(68) \qquad a_3' \equiv a_3 + a_2, \quad b_2' \equiv b_2 + b_3 + a_1 \qquad (\text{mod } 2),$$

induced on the coefficients of F by the transformations (8) and

$$(69) \qquad x_1 = x_1', \quad x_2 = x_2', \quad x_3 = x_3' + x_2'.$$

15. *A fundamental system of invariants of F, under the group Γ generated by the transformations* (8), (51) *and* (69), *is given by* $A, \Delta, J, a_2, b_3, a_1\alpha_2$ *and*

$$(70) \qquad \beta = b_1(b_3 + \alpha_2).$$

It suffices to prove that these seven functions, which are evidently invariant under Γ, completely characterize the classes of forms F under Γ. There are six cases.

(i) $b_3 \equiv a_2 \equiv 1$. Replacing x_1 by $x_1 + a_1x_2$ and x_3 by $x_3 + a_3x_2$, we get

$$F = \beta x_1^2 + \Delta x_2^2 + x_3^2 + x_1x_3.$$

(ii) $b_3 \equiv 1$, $a_2 \equiv 0$, $a_1\alpha_2 \equiv 1$. Replacing x_3 by $x_3 + a_3x_1$, we get

$$F = \Delta x_1^2 + b_2x_2^2 + x_3^2 + x_2x_3.$$

If $\Delta \equiv 0$, then $b_2 \equiv J$. If $\Delta \equiv 1$, we replace x_1 by $x_1 + b_2x_2$ and get

$$x_1^2 + x_3^2 + x_2x_3.$$

(iii) $b_3 \equiv 1$, $a_2 \equiv a_1\alpha_2 \equiv 0$. Replacing x_3 by $x_3 + b_1x_1 + b_2x_2$, we get

$$x_3^2 + \Delta x_1x_2.$$

(iv) $b_3 \equiv 0$, $a_2 \equiv 1$. After replacing x_3 by $x_3 + a_3x_2$, we obtain a form with also $a_3 \equiv 0$. Taking this as F, and replacing x_1 by $x_1 + a_1x_2$, we get

$$b_1x_1^2 + \Delta x_2^2 + x_1x_3.$$

Replacing x_3 by $x_3 + b_1x_1$, we get $\Delta x_2^2 + x_1x_3$.

(v) $b_3 \equiv a_2 \equiv 0$, $a_1\alpha_2 \equiv 1$. Replacing x_3 by $x_3 + a_3x_1 + b_2x_2$, we get

$$\beta x_1^2 + x_2x_3.$$

(vi) $b_3 \equiv a_2 \equiv a_1\alpha_2 \equiv 0$. Then F is the binary form f in (46). The effective part of Γ is now the subgroup Γ_1 generated by (8). Now

$$\beta \equiv b_1, \quad A + 1 \equiv a_3, \quad J \equiv B + (b_1 + 1)\alpha_3, \quad B = b_2(b_1 + \alpha_3).$$

These seminvariants b_1, a_3, B of f completely characterize the classes of forms f under Γ_1. For, if $a_3 \equiv b_1$,

$$f = b_1x_1^2 + Bx_2^2 + b_1x_1x_2;$$

while if $a_3 \equiv b_1 + 1$, we replace x_1 by $x_1 + b_2x_2$ and get

$$b_1x_1^2 + (b_1 + 1)x_1x_2.$$

16. The number of classes of forms F in the respective cases (i)–(vi) is 4, 3, 2, 2, 6. Hence there are exactly 19 linearly independent invariants of F under the group Γ. As these we may take

$$1, \quad a_2, \quad a_1\alpha_2, \quad A, \quad b_3, \quad b_3a_2, \quad b_3a_1\alpha_2, \quad b_3A,$$

$$\Delta = b_1a_1 + \cdots, \quad a_2\Delta = b_1a_1a_2 + \cdots,$$

$$\beta = b_1(b_3 + \alpha_2), \quad a_2\beta = b_1b_3a_2, \tag{71}$$

(71) $A\beta=b_1(b_3+1)A,\quad b_3\Delta=b_1b_3a_1+\cdots,\quad a_2b_3\Delta=b_1b_3a_1a_2+\cdots,$

$$J = b_1b_2b_3 + \cdots,\quad a_2J = b_1b_2b_3a_2 + \cdots,$$

$$b_3J = b_1b_2b_3(a_1a_2 + a_1 + a_2) + \cdots,\quad AJ = b_1b_2b_3A + \cdots.$$

These are linearly independent since the first eight do not involve b_1, while all the terms with the factor b_1 in the next seven are given explicitly, likewise all with the factor $b_1b_2b_3$ in the last four. Hence *the* 19 *functions* (71) *form a complete set of linearly independent invariants of* F *under the group* Γ.

17. Hence, in § 14, S_1 is a linear combination of the functions (71). By (67_2), $S + S_1$ is of the form (57) if S_2 be denoted by (56). Now a_3b_1 occurs in J, AJ, b_3J, a_2J, $A\beta$, but in no further function (71). In the first three, a_3b_1 is multiplied by the linearly independent functions (62), respectively; in the last two by $b_3\alpha_1a_2$ and $\alpha_1\alpha_2(b_3 + 1)$, whose sum is congruent to the first function (62). Hence the part of $S + S_1$ involving J, $\cdots$, $A\beta$ is a linear combination of

$$(72)\qquad (b_3 + a_2)J = b_1b_2b_3a_1\alpha_2 + b_2b_3a_1\alpha_2\alpha_3,$$

$$(73)\qquad J + b_3J + A\beta = (b_3 + 1)(b_1b_2\alpha_1\alpha_2 + b_2A + A).$$

But b_1 occurs in just six of the functions (71) other than the five just considered. Thus the factor pa_1 of b_1 in (57) is a linear combination of the coefficients of b_1 in (72), (73), β, $a_2\beta$, Δ, $a_2\Delta$, $b_3\Delta$, $a_2b_3\Delta$. Now a_1 is a factor of the coefficients of b_1 in all except the second, third and fourth, while in these the coefficients are

$$(b_3 + 1)b_2\alpha_1\alpha_2,\quad b_3 + a_2 + 1,\quad a_2b_3$$

and are linearly independent. Hence (73), β, $a_2\beta$ do not occur in $S + S_1$. By (57), the latter has no constant term and hence involves 1, A only in the combination $A + 1$. This cannot occur since the total coefficient of a_3 must be of the form pR and hence vanish for $b_3 \equiv a_2$. At the same time we see that the sum of the constant multipliers of Δ, $a_2\Delta$, $b_3\Delta$, $a_2b_3\Delta$ is zero modulo 2. Hence $S + S_1$ is a linear combination of the functions

a_2, b_3, b_3a_2, $a_1\alpha_2$, and the last six in (74) below. Like (57), this combination must vanish for $a_1 \equiv 0$, $b_3 \equiv a_2$. Since all but the first three of the ten functions then vanish, the sum of the multipliers of these three must be zero modulo 2. Hence $S + S_1$ is a linear combination of

$$(74)\qquad \begin{array}{l} b_3 + a_2, \quad a_2(b_3+1), \quad a_1\alpha_2, \quad b_3a_1\alpha_2, \quad b_3A, \\ \Delta\alpha_2, \quad \Delta(b_3+1), \quad \Delta(a_2b_3+1), \quad (b_3+a_2)J. \end{array}$$

18. Without altering the invariant S, we may simplify S_1 by subtracting from C constant multiples of $L^{4n-1}K$ and its product by Δ, where K is given by (29), and hence delete $a_2(b_3+1)$ and $\Delta(a_2b_3+1)$ from the terms (74) of S_1. Then

$$S_1 = S + m\Delta\alpha_2 + m_1\Delta(b_3+1) + m_2(b_3+a_2)J$$
$$+ m_3(b_3+a_2) + m_4a_1\alpha_2 + m_5b_3a_1\alpha_2 + m_6b_3A.$$

The coefficient T of $x_3^{\omega-1}x_2$ in C is obtained from S_1 by applying the substitution $(a_1a_2)(b_1b_2)$ induced by (x_1x_2). In view of the transformation (8), we see that $T' = T + S_1$, where T' is derived from T by (22). Hence

$$S = (m+m_1)\Delta + m_1b_3\Delta + m_2b_3J$$
$$+ (m_4+m_5b_3)(a_1a_2+a_1+a_2) + m_3b_3 + m_6b_3A.$$

Let Σ be the sum of the second member and the function obtained from it by the substitution $(a_2a_3)(b_2b_3)$. Thus $\Sigma \equiv 0$. Taking $b_3 \equiv b_2$, we get $m_4 \equiv m_5 \equiv 0$. Then

$$\Sigma = (b_2+b_3)I, \quad I = m_1\Delta + m_2J + m_3 + m_6A.$$

Applying to Σ the substitution (68), we get $(b_2+a_1)I = 0$. Applying $(a_1a_3)(b_1b_3)$ to the latter, we get $(b_2+a_3)I = 0$. Adding, we get $(a_1+a_3)I = 0$. Applying (50), we see that $a_3I = 0$. Then each $a_iI = 0$, so that $I = gA$, where g is a constant. By $\Sigma = 0$, $g = 0$. Thus m_1, m_2, m_3, m_6 are zero. Hence $S = m\Delta$, $S_1 = m\Delta a_2$. But

$$(75)\qquad E = F(L^4 + \Delta F^2) + (\Delta + A)L^6 = \Delta x_3^6 + \cdots.$$

Hence $C - Q_2^{n-1}E$ has the leader zero. *Any covariant of order* $\omega = 4n + 2 > 2$ *differs from a linear combination of the covariants* (65) *and* $Q_2^{n-1}E$ *by a covariant whose leader is zero.*

19. *Regular and Irregular Covariants; Rank.*—A covariant shall be called regular or irregular according as it has not or has the factor L_3, given by (43). The quotient of an irregular covariant by L_3 is a covariant. Hence the determination of all irregular covariants reduces to that of the regular covariants. If a covariant has a linear factor it has as a factor each of the seven ternary linear functions incongruent modulo 2, whose product is L_3. Hence a regular covariant has a non-vanishing component involving only x_1, x_3. In a regular covariant C without terms x_i^ω (i. e., with leader zero), this component has the factors x_1, x_3 and (by the covariant property) also $x_1 + x_3$. The product of these three linear factors was denoted by ξ_2 in (40). Let ξ_2^m be the highest power of ξ_2 which is a factor of the component and let n be the degree of the quotient in the x's. Then C may be given the notation

$$(76) \qquad R_{m,\,n} = \sum_{i=1}^{3} f_i \xi_i^m + x_1 x_2 x_3 \phi,$$

where, if $n = 0$, f_2 is a function of the a's and b's not identically zero, while, if $n > 0$, f_2 is a function also of x_1, x_3 in which the coefficients of x_1^n and x_3^n are not zero; f_1 is a function of x_2, x_3; f_3 of x_1, x_2.

The regular covariant (76) shall be said to be of *rank m*. In an irregular covariant the component free of x_2 is zero and hence is divisible by an arbitrary power of ξ_2; it is proper and convenient to say that an irregular covariant is of infinite rank.

Any covariant of rank zero differs from one of rank greater than zero by a polynomial in the known covariants

$$(77) \qquad A, \quad \Delta, \quad J, \quad F, \quad L, \quad Q_2.$$

This is a consequence of the theorems in §§ 11–18, where the polynomial is given explicitly. Any product, of order ω in the

x's, of powers of the covariants (77) can be reduced by means of the syzygies

$$JL = 0, \quad AL^2 = AF, \quad (\Delta + A + J + 1)(FL + K) = 0,$$

$$(78) \quad AK = 0, \quad FL^2 + (A + \Delta)L^4 + \Delta F^2 + \Delta Q_2 = LK,$$

$$F^3 + Q_2F = L^3K + (\Delta + J)K^2 + (\Delta + 1)LG + (A + 1)Q_1,$$

to a sum of covariants of order ω given in §§ 11–18 and a linear function, with covariant coefficients, of K, Q_1 and

$$G = Q_2L + L^5 = \Sigma\xi_2[\beta_3(\beta_1 + 1)x_3^2 + (\beta_1\beta_3 + 1)x_3x_1$$

$$(79) \quad + \beta_1(\beta_3 + 1)x_1^2] + x_1x_2x_3[(\beta_1 + \beta_2 + \beta_3 + 1)$$

$$\times (x_1x_2 + x_1x_3 + x_2x_3) + \Sigma(\beta_i + 1)x_i^2].$$

Here G and K, given by (42), are of rank 1, while $Q_1 = \xi_2^2 + x_2$ ( ) is of rank 2. As this theorem is not presupposed in what follows, its proof is omitted. However, it led naturally to the important relations (75) and (79) and showed that no new combinations of the covariants (77) of rank zero yield covariants of rank > 0, a fact used as a guide in the investigation of the latter covariants.

Regular Covariants R_{m0}, §§ 20–22

20. A separate treatment is necessary for covariants (76) with $n = 0$. Then each f_i is a function of the coefficients a_j, b_j. Since the factor ξ_3^m of the part $f_3\xi_3^m$ of R_{m0} free of x_3 is unaltered by every linear transformation on x_1 and x_2, f_3 is a linear combination of the functions (49) and their products by b_3. Also, f_3 must be unaltered by

$$(80) \quad x_1 = x_1' + x_3': \quad a_1' = a_1 + a_3, \quad b_3' = b_3 + b_1 + a_2.$$

Both conditions are evidently satisfied by the ternary invariants and by a_3 and q, in (47). In view of (53), we may employ

$$AJ, \quad J, \quad a_3\Delta, \quad \Delta, \quad a_3J, \quad qA, \quad A$$

to replace in turn

$$b_3\alpha_1\alpha_2a_3q, \quad b_3\alpha_1\alpha_2a_3, \quad a_3j, \quad j, \quad a_3b_3q, \quad \alpha_1\alpha_2a_3q, \quad \alpha_1\alpha_2a_3,$$

since a term previously replaced is not introduced later. Thus f_3 is a linear combination of these seven functions, a_3, q, a_3q, and

$$\alpha_1\alpha_2,\quad \alpha_1\alpha_2 q,\quad b_3,\quad b_3a_3,\quad b_3q,\quad b_3\alpha_1\alpha_2,\quad b_3\alpha_1\alpha_2 q,\quad b_3j,\quad b_3a_3j.$$

Give to any linear function $m_1\alpha_1\alpha_2 + \cdots$ of these the notation

$$\sigma = \alpha a_1 b_3 + \beta a_1 + \gamma b_3 + \delta.$$

Call e the increment $b_1 + a_2$ to b_3 in (80) and employ e to eliminate b_1. Then σ is unaltered by (80) if and only if

$$\alpha e \equiv 0,\quad \alpha a_3 \equiv 0,\quad \beta a_3 \equiv \gamma e \qquad (\text{mod } 2).$$

Since b_3 does not occur in q or j, nor a_1 in q, we have

$$\alpha = m_6\alpha_2 + m_7\alpha_2 q + m_8(e + a_2 + a_3) + m_9 a_3(e + a_2 + a_3).$$

Thus $\alpha e \equiv 0$ gives $m_6 \equiv m_7 \equiv 0$, $m_8 \equiv m_9$. Then $\alpha a_3 \equiv 0$ gives $m_9 \equiv 0$. Now

$$\beta = m_1\alpha_2 + m_2\alpha_2 q,\quad \gamma = m_3 + m_4 a_3 + m_5 q,$$

and $\beta a_3 \equiv \gamma e$ readily gives $\sigma \equiv 0$. *Any function of* b_3 *and the invariants* (49) *of* f *and* l, *which is unaltered by* (80), *is a linear combination of the ternary invariants* (45) *and* a_3, q, a_3q, $a_3\Delta$, a_3J, qA.

21. For $n = 0$ and m even, there exists a covariant (76) in which f_3 is any function specified in the preceding theorem. For, if I is any ternary invariant, $IQ_1^{m/2}$ has $f_3 = I$. By (42) and (41), K^m and $W^{m/2}$ are of the form (76) with $f_3 = a_3$ and $\beta_1\beta_2 + 1$, respectively; they may be multiplied by any invariant. By (19) and (47), we have

$$(81)\quad \beta_1\beta_2 + 1 = q + \alpha_3\Delta + A + 1,\quad a_3q = \alpha_3\Delta + q\Delta + a_3J.$$

Hence we obtain q, then qA, $q\Delta$, and therefore a_3q. *Any covariant with* $n = 0$, m *even, differs by an irregular covariant from a linear function of*

$$IQ_1^{m/2}, I_1K^m, I_2W^{m/2}\quad (I=1,\Delta,A,J,AJ;\ I_1=1,\Delta,J;\ I_2=1,\Delta,A).$$

22. For $n = 0$ and m odd, we may delete the terms a_3I_1 from f_3 by use of I_1K^m. First, let $m = 1$ and apply transformation (51); we get

$$\xi_1 = \xi_1' + \xi_3', \quad \xi_2 = \xi_2', \quad \xi_3 = \xi_3',$$
$$(82) \qquad R' = f_1\xi_1' + f_2\xi_2' + (f_1 + f_3)\xi_3' + (x_1'x_2'x_3' + x_1'^2x_2')\phi.$$

Thus $\phi = 0$. Since $f_3 = I + I_2q$, condition $f_1 + f_3 = f_3'$ gives

$$I = I_2(a_1b_1 + \alpha_2b_2 + a_3b_3 + a_2\alpha_3 + a_1a_2).$$

Add to this the relation obtained by permuting the subscripts 1, 2. Thus

$$0 = I_2(b_1 + b_2 + a_2\alpha_3 + a_1\alpha_3).$$

The increment under (22) is $I_2(b_1 + a_3 + a_2\alpha_3) = 0$. Now I_2 is of the form $x + y\Delta + zA$, where x, y, z are constants. From the terms in b_1b_2, we get $y = 0$. Then $x = z = 0$. *The only covariants are therefore* I_1K.

Second, let $m > 1$. Then $KW^{(m-1)/2}$ is of the form (76) with $f_3 = a_3q + a_3$, by (81_1). Hence we may set

$$f_3 = I + cq + dqA \qquad (c, d \text{ constants}).$$

In R given by (76), let g denote the coefficient of

$$(83) \qquad x_1x_2x_3 \cdot x_2^m x_3^{2m-3}.$$

In the function derived from R by the transformation (51), the term corresponding to (83) has the coefficient $g + f_1$, since by (82) the ξ_i parts contribute only one such term, that from $f_1\xi_1'^{m-1}\xi_3'$. Now

$$f_1 = I + cq' + dq'A, \quad q' = b_2b_3 + (b_2 + b_3)\alpha_1.$$

When g is given the notation (56), $g' - g = f_1$ is the function (57). But a_3b_1 occurs in f_1 only in J and AJ and in them with the linearly independent multipliers (62). Hence

$$I = n_1(A + 1) + n_2\Delta.$$

The coefficient of a_3 in f_1 is now

$$n_1\alpha_1\alpha_2 + n_2(a_1a_2 + b_3) + dq'\alpha_1\alpha_2 = p(b_3 + a_2).$$

Taking $b_3 \equiv a_2$, we see that $n_1 \equiv n_2 \equiv d \equiv 0$. Thus $f_1 = cq'$. By (57) for $a_1 \equiv 0$, $b_3 \equiv a_2$, we get $c \equiv 0$. *Any covariant with $n = 0$ and m odd differs by an irregular covariant from a linear function of K^m, ΔK^m, JK^m and, if $m > 1$, $KW^{(m-1)/2}$.*

Covariants of Rank Unity, §§ 23–26

23. Henceforth let $m > 0$, $n > 0$ in (76) and set

(84) $$f_2 = Sx_3^n + S_1x_3^{n-1}x_1 + S_2x_3^{n-2}x_1^2 + \cdots \quad (S \neq 0).$$

Since S is unaltered by the group Γ of § 15, it is a linear combination of the functions (71). We may omit the functions $a_2(b_3+1)$ and $\Delta a_2(b_3+1)$, since K^mL^n is of the form (76) with $S = a_2(b_3+1)$. Thus

(85) $$S = I + a_2I_1 + b_3I_2 + k_1a_1\alpha_2 + k_2b_3a_1\alpha_2 + k_3\beta + k_4a_2\beta + k_5A\beta,$$

where I is any invariant, I_1 a linear function of $1, \Delta, J$; I_2 one of $1, A, \Delta, J$; while $\beta = b_1(b_3+\alpha_2)$.

First, let $m = 1$. If T and B are the coefficients of x_2^n in f_3 and f_1, transformation (51) replaces the covariant (76) by a function in which, by (82), the coefficient of $x_1'x_2'^{n+2}$ is

(86) $$T + B = T',$$

where T' is derived from T by the induced substitution (50). But T is obtained from S by the interchange [23] of subscripts, and B from T by [13]. We thus find by (86) that

$$I = b_2I_2 + (k_1 + k_2b_2)(a_1 + a_3\alpha_1)$$
$$+ k_3(a_1b_1 + a_2b_2 + a_3b_3 + a_1a_2 + a_2\alpha_3)$$
$$+ k_4b_2(a_1b_1 + a_3b_3 + a_1a_2 + a_2a_3).$$

Let Σ be the sum of the second member and the function obtained by applying $(a_2a_3)(b_2b_3)$ to it. In $\Sigma = 0$, set $b_2 = b_3$; we get

$$\{k_1 + k_3 + b_3(k_2 + k_4)\}(a_2 + a_3)\alpha_1 = 0, \quad k_3 \equiv k_1, \quad k_4 \equiv k_2.$$

Then $\Sigma = 0$ may be written in the form

$$(b_2 + b_3)\lambda = 0, \quad \lambda = I_2 + k_2(\Delta + A + 1).$$

As in § 18, $\lambda = 0$. Thus I_2 and I are the products of $\Delta + A + 1$ by k_2, k_1, so that

$$S = (k_1 + k_2b_3)(\Delta + A + 1) + a_2I_1 + k_1(b_1b_3 + b_1\alpha_2 + a_1\alpha_2) \tag{87}$$
$$+ k_2b_3(a_2b_1 + a_1\alpha_2) + k_5A(b_1b_3 + b_1).$$

For n odd, S is the increment to S_1 under (50) and hence has no term containing a_3b_1. If t is the coefficient of J in I_1, a_3b_1 occurs in (87) only in ta_2J and in the final part, being multiplied by $ta_2\alpha_1b_3$ and $k_5\alpha_1\alpha_2(b_3 + 1)$, respectively. Hence $t \equiv k_5 \equiv 0$. Since S is of the form (57), the coefficient of b_1 must vanish if $a_1 \equiv 0$. Thus

$$k_1(b_3 + \alpha_2) + k_2b_3a_2 = 0, \quad k_1 \equiv k_2 \equiv 0.$$

Now $S = a_2I_1 = a_2(u + v\Delta)$ must vanish for $a_1 \equiv 0$, $b_3 \equiv a_2$ by (57); then $\Delta = a_2(b_2 + a_3)$, so that $u = v = 0$, $S = 0$. *Any covariant with $m = 1$ and n odd differs from one of rank > 1 by a linear function of KL^n, ΔKL^n.*

24. For $m = 1$, $n = 4\nu$, we may delete a_2I_1 from (85) by use of $I_1KQ_2^{\nu}$. Set $f_1 = Bx_2^n + \cdots + B_nx_3^n$. Then (51) replaces (76) by

$$R' = \xi_2[Sx_3^n + S_1x_3^{n-1}x_1 + (S_1 + S_2)x_3^{n-2}x_1^2 + \cdots] + \xi_3f_3$$
$$+ (\xi_1 + \xi_3)[B_n(x_3^n + x_3^{n-4}x_1^4 + \cdots)$$
$$+ B_{n-1}x_2(x_3^{n-1} + x_3^{n-2}x_1 + \cdots)] + (x_1x_2x_3 + x_1^2x_2)\phi'.$$

Since S_1 is the increment of S_2, it is a linear combination of the functions (74). By use of $L^{n-3}Q_1$, $L^{n-3}K^2$ and their products by A and Δ, we may, without disturbing S, delete from S_1

$$b_3+\alpha_1\alpha_2+1, \quad Ab_3, \quad \Delta+b_3\Delta+b_3\alpha_1\alpha_2, \quad a_2(b_3+1), \quad a_2\Delta(b_3+1).$$

Hence we may set

$$S_1 = t_1(b_3 + a_2) + t_2b_3a_1\alpha_2 + t_3\Delta\alpha_2 + t_4(b_3 + a_2)J.$$

Applying $(a_1a_2)(b_1b_2)$ to S and S_1, we obtain B_n and B_{n-1}. Let l be the coefficient of $x_2x_3^{n-1}$ in ϕ. By the coefficient of

$x_1x_2x_3 \cdot x_2x_3^{n-1}$ in R', we have

$$B_n + B_{n-1} + l = l'.$$

For l given by (56), $B_n + B_{n-1}$ is given by (57). By the coefficient of a_3b_1, we get $t_4 \equiv 0$. The coefficient of a_3 must vanish for $b_3 = a_2$. Hence

$$k_1\alpha_1 + (k_2 + t_3)\alpha_1 a_2 + k_5\alpha_1\alpha_2 b_2 = 0, \quad k_1 \equiv k_5 \equiv 0, \quad t_3 \equiv k_2,$$

$$S = k_2b_3(\Delta + A + 1 + a_1\alpha_2 + a_2b_1).$$

The coefficient of k_2 equals that of $\xi_2x_3^n$ in

$$GFQ_2^{\nu-1} + \Delta KL^n + \Delta KQ_2^{\nu}.$$

Any covariant with $m = 1$, $n = 4\nu$, *differs from one with* $m \geqq 2$ *by a linear function of* KL^n, ΔKL^n, IKQ_2^{ν}, $GFQ_2^{\nu-1}$ $(I = 1, \Delta, J)$.

25. For $m = 1$, $n = 4\nu + 2$, we may delete a_2I_1 from S, given by (87), by use of $I_1Q_2^{\nu}M$. The coefficient of $\xi_2x_3^n$ in $Q_2^{\nu}G$ is

$$d = \beta_3(\beta_1 + 1) = A + (b_1 + 1)(\alpha_1\alpha_2 + b_3) + b_3\alpha_2\alpha_3.$$

The coefficient of k_1 in S equals $d + a_2\Delta + a_2(b_3 + 1)$, the final term of which was reached in § 23, and $a_2\Delta$ above. The coefficients of k_5 and k_2 in S equal Ad and

$$b_1b_3(a_1+a_2)+b_3(a_2b_2+a_1a_3+a_2a_3+a_2)=\Delta d+a_2(J+1)+a_2(b_3+1),$$

respectively. *Any covariant with* $m = 1$, $n = 4\nu + 2$, *differs from one with* $m \geqq 2$ *by a linear function of* KL^n, ΔKL^n, $IQ_2^{\nu}G$, $I_1Q_2^{\nu}M$ $(I = 1, A, \Delta;\ I_1 = 1, \Delta, J)$.

For use in § 26, we replace $Q_2^{\nu}M$ by $Q_2^{\nu}FK$, noting that

$$M = (F + L^2)K \tag{88}$$

and that $Q^{\nu}L^2K$ differs from KL^n by a covariant of rank 2.

26. By the last four theorems, any covariant of rank 1 differs from one of rank $\geqq 2$ by $CK + DG$, where C and D are known covariants of rank zero. Taking as C_1 and D_1 arbitrary func-

tions of the proper degree in the x's, of the generators (77) of covariants of rank zero, I found the syzygies needed to reduce $C_1K + D_1G$ to an expression differing from the above $CK + DG$ by a covariant of rank $\geqq 2$, in which those of rank 2 are linear combinations of K^2, KG, G^2, W, Q_1 and the new one

$$V = GF^2 + \Delta Q_2G + (\Delta + J + 1)Q_2FK + \Delta L^3K^2 + \Delta L^3Q_1 = \xi_2^2x_3^3v + \cdots, \tag{89}$$

where

$$v = a_2 + b_3(1 + a_1\alpha_2). \tag{90}$$

The only new syzygies needed for this reduction are

$$\begin{gathered} LG \equiv Q_2L^2 + L^6 = W, \quad FLK = \Delta W + \Delta Q_1 + (J + 1)K^2, \\ (F^2 + L^4 + Q_2)K = (A + 1)L_3, \\ (\Delta + 1)(FG + KL^4 + KQ_2) + JKQ_2 = ALQ_1 + \omega L_3, \end{gathered} \tag{91}$$

in which ω is an invariant not computed. Proof need not be given of these facts since we presuppose below merely the existence of relation (89) which may be verified independently. Of course, the fact that V is the only new covariant of rank 2 was a guide in the later investigation.

Covariants of Even Rank $m = 2\mu > 0$, §§ 27–29

27. First, let n be odd. In the covariant (76) replace x_3 by $x_3 + x_1$. In view of (82), we get

$$R' = f_2'\xi_2^m + f_3\xi_3^m + f_1'(\xi_1^2 + \xi_3^2)^\mu + (x_1x_2x_3 + x_1^2x_2)\phi'.$$

Using the notation (84) for f_2, we have $S_1' = S_1 + S$ in f_2'. Thus, as in § 17, S is a linear combination of the functions (74). Now $Q_1^\mu L^n$ and its products by A and $A + \Delta$ are covariants (76) with S given by (63). Using also K^mL^n, in which $S = a_2(b_3+1)$, and its product by Δ, we may set

$$S = k_1(b_3 + a_2) + k_2b_3a_1\alpha_2 + k_3\Delta\alpha_2 + k_4(b_3 + a_2)J.$$

In $x_1x_2x_3\phi$, let g be the coefficient of

$$x_1x_2x_3 \cdot x_2^{2\mu-1}x_3^{4\mu+n-2} = (x_2^2x_3^4)^\mu x_3^{n-1}x_1.$$

Such a term occurs in neither of the first two parts of R', since they are functions of only two variables. To obtain such a term from the third part of R', we must omit terms with the factor ξ_3^2 (and hence x_1^2) and take $(x_2x_3^2)^{2\mu}$ in $\xi_1^{2\mu}$, so as not to make the degree in x_2 too high. Hence if T be the coefficient of x_3^n in f_1, $g' \equiv g + T$. Now $(a_1a_2)(b_1b_2)$ replaces S by T. The resulting T must be of the form (57). By the coefficient of a_3b_1, $k_4 \equiv 0$; cf. (72). By the coefficient $k_3\alpha_1b_3$ of a_3, $k_3 \equiv 0$. Since $T \equiv 0$ for $a_1 \equiv 0$, $b_3 \equiv a_2$, we get $k_1 \equiv k_2$. Hence $S = k_1v$, where v is given by (90).

For $n = 1$, $f_2 = Sx_3 + S_1x_1$. Thus $S_1 = k_1v'$, where v' is derived from v by interchanging the subscripts 1 and 3. Then $S_1' \equiv S_1 + S$ gives $k_1 \equiv 0$.

For $n \geqq 3$, $Q_1^{\mu-1}L^{n-3}V$ is of the form (76) with $S = v$, since $\beta_3v = 0$.

Any covariant with n odd, $m = 2\mu > 0$, differs from one of rank $> m$ by a linear combination of $IQ_1^\mu L^n$ $(I = 1, A, \Delta)$, K^mL^n, ΔK^mL^n and, if $n > 1$, $Q_1^{\mu-1}L^{n-3}V$.

28. For $m = 2\mu > 0$, $n = 4\nu > 0$, the coefficients of $\xi_2^mx_3^n$ in

$$
\begin{gathered}
Q_1^\mu Q_2^\mu, \quad K^mQ_2^\nu, \quad Q_1^\mu F^{2\nu}, \quad Q_1^\mu L^n, \quad K^mL^n, \\
Q_1^{\mu-1}Q_2^{\nu-1}G^2, \quad K^{m-2}Q_2^{\nu-1}G^2
\end{gathered}
\tag{92}
$$

are respectively

$$1, \quad a_2, \quad b_3, \quad \beta_3 + 1, \quad a_2(b_3 + 1), \quad d = \beta_3(\beta_3 + 1), \quad a_2d.$$

These may be multiplied by any invariant. Now

$$\beta_3 + 1 + a_2 + b_3 = a_1\alpha_2,$$

$$\Delta(\beta_3 + 1) + (\Delta + A + 1)b_3 + b_3a_2 + \Delta = b_3a_1\alpha_2,$$

$$d + A + \beta_3 + \alpha_2(\Delta + b_3) = b_1(b_3 + \alpha_2) \equiv \beta,$$

$$a_2d + a_2b_3 = a_2b_1b_3 = a_2\beta, \quad Ad = Ab_1(b_3 + 1) = A\beta.$$

Hence we have a covariant (76) in which the coefficient of $\xi_2^mx_3^n$ is any linear combination of the functions (71). Hence *the covariant differs from one of rank* $> m$ *by a linear function of the covariants* (92), *the products of the first three by any invariant except* 1, *the products of the fourth and fifth by* Δ *and the product of the sixth by* A.

29. For $m = 2\mu > 0$, $n = 4\nu + 2$, the coefficients of $\xi_2^m x_3^n$ in

$$(93) \qquad MK^{m-1}Q_2^{\nu}, \quad K^m L^n, \quad GK^{m-1}Q_2^{\nu}, \quad F^{n/2}Q_1^{\mu}, \quad L^n Q_1^{\mu}$$

are respectively

$$a_2, \quad a_2(b_3 + 1), \quad a_2 b_3(b_1 + 1), \quad b_3, \quad b_3 + \alpha_1\alpha_2 + 1.$$

Linear combinations of products of these by invariants give*

$$a_2, \quad a_2\Delta, \quad a_2 J, \quad a_2 b_3, \quad \Delta a_2 b_3, \quad a_2 b_1 b_3, \quad I b_3, \quad a_1\alpha_2, \quad \Delta + b_3 a_1 \alpha_2.$$

Since S and S_1 are unaltered by the group Γ of § 15, they are linear combinations of the functions (71). Deleting the above functions a_2, $a_2\Delta$, $\cdots$ from S, we have

$$S = I + c\beta + eA\beta, \quad \beta = b_1(b_3 + \alpha_2),$$

where c and e are constants, and I is an invariant. Set

$$f_1 = Bx_2^n + B_1 x_2^{n-1}x_3 + \cdots + B_{n-1}x_2 x_3^{n-1} + B_n x_3^n,$$

and call σ the coefficient of

$$(94) \qquad x_1 x_2 x_3 \cdot x_2^{4\mu+n-2}x_3^{2\mu-1} = (x_2^2 x_3)^{2\mu} x_2^{n-1} x_1$$

in $x_1 x_2 x_3 \phi$. The coefficient† of (94) in R' of § 27 is $B_1 + \sigma$. Hence

$$\sigma' - \sigma = B_1,$$

if (50) replaces σ by σ'. Thus B_1 must be of the form (57).

For $n = 2$, S_2 is derived from S by applying $(a_1 a_3)(b_1 b_3)$. Then (67_2) gives S_1. Applying $(a_1 a_2)(b_1 b_2)$ to S_1, we get

$$B_1 = I + c(b_2 b_3 + b_2\alpha_1 + b_3\alpha_1) + eA(b_2 b_3 + b_2 + b_3).$$

* For the last two, use the first two of the four equations in § 28.

† The first part of R' is free of x_2, the second of x_3, while in the third part ξ_3^2 has the factor x_1^2, and in $f_1'\xi_1^{2\mu}$ there is a single term (94) and it has the coefficient B_1.

Since this must be of the form (57), we get $I = 0$, $c = e = 0$. *A covariant with* $m = 2\mu$, $n = 2$, *differs from one of rank* $> m$ *by a linear function of*

$$iMK^{m-1},\quad K^mL^2,\quad \Delta K^mL^2,\quad GK^{m-1},\quad IFQ_1^{\mu},\quad L^2Q_1^{\mu},\quad \Delta L^2Q_1^{\mu}$$
$$(i = 1,\ \Delta,\ J;\ I = 1,\ A,\ \Delta,\ J).$$

For $n > 2$, we may delete Δ from the part I of S by use of $EQ_1^{\mu}Q_2^{\nu-1}$, where E is given by (75). Without disturbing S we may delete $a_2(b_3 + 1)$ and its product by Δ from S_1 by use of $K^{2\mu+1}L^{n-3}$, since the term of $\xi_2^{2\mu}f_2$ with the coefficient S_1 is the term of highest degree in x_3 in $\xi_2^{2\mu+1}(S_1x_3^{n-3} + \cdots)$. Since $S + S_1$ is a linear combination of the functions (74),

$$\text{(95)}\qquad \begin{aligned} S_1 = S + t_1(b_3 + a_2) + t_2a_1\alpha_2 + t_3b_3a_1\alpha_2 + t_4b_3A + t_5\Delta\alpha_2 \\ + t\Delta(b_3 + 1) + t_0(b_3 + a_2)J. \end{aligned}$$

Apply $(a_1a_2a_3)(b_1b_2b_3)$ to B_1, of the form (57). Hence

$$\text{(96)}\qquad S_1 = p\rho a_1 + pa_2b_2 + pa_2\rho + ra_2 + s\rho,\quad \rho = b_1 + a_3.$$

Now a_1b_2 occurs in S only in the terms J, AJ of I and in the part of (95) after S only in the last term, given by (72). In these the factors of a_1b_2 are linearly independent. Hence $t_0 = 0$, $I = x(A + 1)$. The coefficient of a_1 in S_1 must vanish for $b_1 \equiv a_3$, and S_1 itself if also $a_2 \equiv 0$. Hence

$$c = t_2 = x,\quad t_1 = t_3 = t_4 = t,\quad t_5 = x + t,$$

$$\begin{aligned} S_1 = x(A + 1 + b_1b_3 + b_1\alpha_2 + a_1\alpha_2 + \Delta\alpha_2) + eAb_1(b_3 + 1) \\ + t(b_3 + a_2 + b_3a_1\alpha_2 + b_3A + b_3\Delta + a_2\Delta). \end{aligned}$$

Call ϵ the coefficient in $x_1x_2x_3\phi$ of

$$x_1x_2x_3 \cdot x_2^{2\mu}x_3^{4\mu+n-3} = (x_2^2x_3^4)^{\mu}x_1x_2x_3^{n-2}.$$

In R' of § 27, the coefficient of this product is $\epsilon + B_{n-1}$. Hence B_{n-1} is of the form (57). Interchanging the subscripts 1 and 2 in B_{n-1}, we get S_1. Thus the coefficient of a_3 in S_1 vanishes for $b_3 = a_1$. Hence $S = S_1 = 0$. *Any covariant with* $n > 2$ *differs from one of rank* $>m$ *by a linear combination of*

$$iMK^{m-1}Q_2^{\nu},\ jK^mL^n,\ GK^{m-1}Q_2^{\nu},\ IF^{n/2}Q_1^{\mu},\ jL^nQ_1^{\mu},\ EQ_1^{\mu}Q_2^{\nu-1}$$
$$(i=1,\ \Delta,\ J;\ j=1,\ \Delta;\ I=1,\ A,\ \Delta,\ J).$$

COVARIANTS OF ODD RANK $m=2\mu+1>1$, §§ 30–31

30. Replacing x_3 by x_3+x_1 in the covariant (76), we get

$$R'=f_2'\xi_2^m+f_3\xi_3^m+f_1'(\xi_1+\xi_3)^m+(x_1x_2x_3+x_1^2x_2)\phi'.$$

In $x_1x_2x_3\phi$, let g be the coefficient of $(x_1x_2^2)(x_2^2x_3)^{m-1}x_2^n$. The coefficient of the corresponding term of R' is $g'\equiv g+B$, where B is that of x_2^n in f_1. Hence B is of the form (57).

First, let n be odd. Then $S_1'=S_1+S$ under (50), so that S is a linear combination of functions (74) with $a_2(b_3+1)$ and its product by Δ deleted (§ 23). Thus S is the sum of the terms (95) after the first. Applying $(a_1a_2a_3)(b_1b_2b_3)$ to B, of the form (57), we see that S is of the form (96). By these two results,

$$S=t(b_3+a_2+b_3a_1\alpha_2+b_3A+b_3\Delta+a_2\Delta).$$

If l is the coefficient of $(x_2x_3^2)^mx_3^{n-1}x_1$ in $x_1x_2x_3\phi$, that in R' is $l'=l+nB_n$. Hence, for n odd, B_n is of the form (57). Interchanging the subscripts 1, 2 in B_n, we get S. Thus the coefficient of a_3 in S vanishes for $b_3\equiv a_1$, so that $t\equiv 0$. *Any covariant with* m *and* n *odd differs from one of rank* $>m$ *by a linear function of* K^mL^n *and* ΔK^mL^n.

31. Finally, let m be odd and n even. According as $n=4\nu$ or $4\nu+2$, $K^mQ_2^{\nu}$ or $K^{m-1}MQ_2^{\nu}$ is of the form (76) with a_2 as the coefficient of $\xi_2^mx_3^n$. Hence we may delete the terms a_2I_1 in (85) and hence the terms a_1I_1 in B of § 23. But (§ 30), B is of the form (57). Now a_3b_1 occurs in J and AJ of I and in b_2J of b_2I_2, having in these linearly independent multipliers. Hence

$$I=x(A+1)+y\Delta,\quad I_2=e+fA+g\Delta.$$

Since the coefficient of a_3 in B shall vanish for $b_3=a_2$, and B itself if also $a_1=0$, we get $k_1=x=y=k_3$, $k_2=f=g=e$.

Thus

$$(97)\quad \begin{aligned} S = x(A + 1 + \Delta + a_1\alpha_2 + b_1b_3 + b_1\alpha_2) + k_4a_2b_1b_3 \\ + g(A + 1 + \Delta + a_1\alpha_2)b_3 + k_5Ab_1(b_3 + 1). \end{aligned}$$

First, let $n = 4\nu + 2$ and write $2\mu + 1$ for m. Then

$$GQ_1{}^{\mu}Q_2{}^{\nu}, \quad K^2GQ_1{}^{\mu-1}Q_2{}^{\nu}$$

have $d = \beta_3(\beta_1 + 1)$ and a_2d as the coefficients of $\xi_2{}^m x_3{}^n$. As in § 25, the coefficients of x, k_4, g, k_5 in (97) equal respectively

$$d + a_2(\Delta + b_3 + 1), \quad a_2d + a_2b_3, \quad \Delta d + a_2d + a_2J, \quad Ad.$$

The terms not containing d are combinations of the above a_2I_1 and $a_2(b_3 + 1)$ of § 23. *Any covariant with $m = 2\mu + 1 > 1$, $n = 4\nu + 2$, differs from one of rank $> m$ by a linear function of*

$$iK^mL^n, \quad I_1K^{m-1}MQ_2{}^{\nu}, \quad IGQ_1{}^{\mu}Q_2{}^{\nu}, \quad K^2GQ_1{}^{\mu-1}Q_2{}^{\nu}$$
$$(i = 1, \Delta;\ I_1 = 1, \Delta, J;\ I = 1, A, \Delta).$$

Next, let $n = 4\nu > 0$. In the last two covariants of the theorem below, the coefficients of $\xi_2{}^{2\mu+1}x_3{}^{4\nu}$ are $a_2b_3(b_1 + 1)$ and $\delta = b_3\beta_3(\beta_1 + 1)$. We had reached covariants in which the corresponding coefficients are a_2I and $a_2(b_3 + 1)I$. Thus we obtain the coefficient of k_4 in (97) and $\delta + \Delta a_2b_3 + a_2b_1b_3$, which equals the coefficient of g. We may therefore set $k_4 = g = 0$. Subtracting covariants of the fourth and fifth types in the theorem, we may take as S_1 the function in § 24, without disturbing S. Applying $(a_1a_2)(b_1b_2)$ to S and S_1, we get B_n and B_{n-1}. If l is the coefficient of $x_1x_2{}^{m+1}x_3{}^{2m+n-2}$ in $x_1x_2x_3\phi$, its coefficient in R' of § 30 is $l' = l + B_n + B_{n-1}$. Thus $B_n + B_{n-1}$ is of the form (57). By the coefficient of a_3b_1, $t_4 = 0$. Since the coefficient of a_3 is zero for $b_3 = a_2$, we get $x = k_5 = t_3 = 0$. Thus $S = 0$. *Any covariant with $m = 2\mu + 1 > 1$, $n = 4\nu > 0$, differs from one of rank $> m$ by a linear function of*

$$K^mL^n, \ \Delta K^mL^n, \ IK^mQ_2{}^{\nu}, \ iL^{n-3}Q_1{}^{\mu+1}, \ iL^{n-3}K^{2\mu+2}, \ G^2KQ_2{}^{\nu-1}Q_1{}^{\mu-1},$$
$$FGQ_2{}^{\nu-1}Q_1{}^{\mu} \quad (i = 1, A, \Delta;\ I = 1, \Delta, J).$$

32. We have now completed the proof of the theorem:

As a fundamental system of modular covariants of the ternary quadratic form F with integral coefficients modulo 2, we may take F, its invariants A, Δ, J, its linear covariant L, its "polar" cubic covariant K, and the universal covariants Q_1, Q_2, L_3.

Incidentally, we have obtained a complete set of linearly independent covariants of each order and rank. We might then find a complete set of independent syzygies. Syzygies whose members are covariants of low rank are given in (78), (88), (91).

33. *References on Modular Geometry.*—Other aspects of the modular geometry of quadratic forms modulo 2 and, in particular, applications to theta functions have been considered by Coble.* For a treatment of non-homogeneous quadratic forms in x, y modulo p (p an odd prime), analogous to that of conics in elementary analytic geometry, but employing only real points on the modular locus, see G. Arnoux, Essai de Géométrie analytique modulaire, Paris, 1911. The earlier paper by Veblen and Bussey was cited in § 7. The paper by Mitchell was cited in § 3. Applications of modular geometries have been made by Conwell.†

The problem of coloring a map has been treated from the standpoint of modular geometry by Veblen.‡

* *Transactions of the American Mathematical Society*, vol. 14 (1913), pp. 241–276.

† *Annals of Mathematics*, ser. 2, vol. 11 (1910), pp. 60–76.

‡ *Annals of Mathematics*, ser. 2, vol. 14 (1912), pp. 86–94.

LECTURE V

A THEORY OF PLANE CUBIC CURVES WITH A REAL INFLEXION POINT VALID IN ORDINARY AND IN MODULAR GEOMETRY

1. *Normal Form of Cubic.*—Consider a ternary cubic form $C(x, y, z)$ with coefficients in a field F not having modulus 2 or 3. After applying a linear transformation with coefficients in F and of determinant unity, we may assume that (1, 0, 0) is an inflexion point. In particular, C lacks the term x^3. If it lacks also x^2y and x^2z, its first partial derivatives vanish for $y = z = 0$. But we shall assume that the discriminant of C is not zero. Hence the coefficient of x^2 may be taken as the new variable y. At the inflexion point (1, 0, 0) the tangent $y = 0$ is to be an inflexion tangent, i. e., meet the cubic in a single point. Hence C lacks the term xz^2. Thus

$$C = x^2y + 2x(\alpha y^2 + \beta yz) + \phi(y, z).$$

Replacing x by $x - \alpha y - \beta z$, we see that x^2y is now the only term involving x. If y were a factor, the discriminant would be zero. Hence the term z^3 occurs. Adding a suitable multiple of y to z, we get

$$C = x^2y + gy^3 + hy^2z + \delta z^3 \qquad (\delta \neq 0). \tag{1}$$

2. *The Invariants s and t.*—The Hessian of (1) is

$$H = -\,3\delta x^2z - h^2y^3 + 9\delta gy^2z + 3\delta hyz^2.$$

The sides of an inflexion triangle form a degenerate cubic belonging to the pencil of cubics $kC + H$. The latter has the factor z only when $k = h = 0$ and the factor $y - lz$ only when $kl = 3\delta$ (as shown by the terms in x^2), where k is a root of

$$k^4 + 18\delta hk^2 + 108\delta^2gk - 27\delta^2h^2 = 0.$$

Before considering the factors involving x, we note that the coefficients of this quartic equation are the values which relative invariants of a general cubic assume for the case of our cubic (1). Indeed, a linear transformation of determinant unity which replaces C by a cubic C' must replace H by the Hessian H' of C', and hence replace the inflexion triangle of C given by a root k of the quartic by that inflexion triangle of C' which is given by the same number k. We denote the invariants by*

$$s = -3\delta h, \quad t = -108\delta^2 g. \tag{2}$$

The above quartic now becomes

$$k^4 - 6sk^2 - tk - 3s^2 = 0. \tag{3}$$

The discriminant Δ of C is such that

$$27\Delta = t^2 - 64s^3. \tag{4}$$

There are four distinct roots of (3) since its discriminant is $-27^3\Delta^2$.

Our earlier results are that $kC + H$ has the factor z only when $k = s = 0$ and the factor $y - 3\delta k^{-1}z$ if k is a root $\neq 0$ of (3). It has the factor $x - \tau y - \rho z$ if and only if

$$3\rho^2 = k, \quad 9\delta^2 k\tau^2 = s^2 + tk/12, \qquad k\rho^2 - 6\delta\rho\tau = s,$$
$$6\delta k\rho\tau - 9\delta^2\tau^2 - sk - t/4 = 0.$$

These conditions are satisfied if and only if k is a root of (3) and

$$\rho = k = 0, \quad 36\delta^2\tau^2 = -t \qquad (k = 0),$$
$$3\rho^2 = k, \quad 6\delta k\tau = \rho(k^2 - 3s) \qquad (k \neq 0).$$

3. *The Four Inflexion Triangles.*—First, let $s = 0$. Then $t \neq 0$ by (4). The root $k = 0$ gives the inflexion triangle with the sides

$$z = 0, \quad x = \pm \tau_1 y \qquad (36\delta^2\tau_1^2 = -t). \tag{5}$$

---

* We have $s = -3^4S$, $t = -3^6T$, where S and T, given in Salmon's Higher Plane Curves, p. 189, are the invariants of the general cubic with multinomial coefficients.

Each root of $k^3 = t$ gives an inflexion triangle

$$(6)\qquad y = \frac{3\delta}{k}z, \qquad x = \pm\,\tau_2\left(y + \frac{6\delta}{k}z\right) \quad (3\cdot 36\delta^2\tau_2{}^2 = t).$$

Next, let $s \neq 0$. Each root of (3) gives an inflexion triangle

$$(7)\qquad y = \frac{3\delta}{k}z, \qquad x = \pm\sqrt{\frac{k}{3}}\left(z + \frac{k^2 - 3s}{6\delta k}y\right).$$

4. *The Parameter* δ.—If we multiply x, y, z by ρ, ρ^{-2}, ρ, we obtain from (1) a cubic with δ replaced by $\delta\rho^3$. If F is the field of all complex numbers, the field of all real numbers, or the finite field of the residues of integers modulo $3j + 2$, a prime, every element is the cube of an element of the field [in the third case, $e \equiv (e^{-j})^3$], so that the parameter δ may be taken to be unity. Although we do not use the fact below, it is in place to state here that for all further fields a new invariant is needed to distinguish the classes of cubics (1). Indeed, two cubics (1), with coefficients in F and with the same invariants s and t and discriminants not zero, are equivalent under a linear transformation with coefficients in F and having determinant unity if and only if the ratio of their δ's is the cube of an element of F.

CRITERIA FOR 9, 3 OR 1 REAL INFLEXION POINTS, §§ 5–9

5. *Inflexion Points when* $s = 0$.—Let κ be a fixed root of $k^3 = t$. Let τ_1 and τ_2 be fixed roots of the equations at the end of (5) and (6). Then

$$(\tau_1/\tau_2)^2 = -3 = (1 + 2\omega)^2, \quad \omega^2 + \omega + 1 = 0.$$

Choose ω so that $\tau_1/\tau_2 = 1 + 2\omega$. Denote the lines $z = 0$, $x = \tau_1 y$, $x = -\tau_1 y$ in (5) by L_1, L_2, L_3. For each value of $i = 0, 1, 2$, denote the three lines (6) with $k = \kappa\omega^i$ by L_{1i}, L_{2i}, L_{3i}, that with the lower sign being L_{3i}. Then the 9 inflexion points and the subscripts of the 4 inflexion lines through each are given in the following table:

(8)

$(1, 0, 0)$	$(\tau_2, 1, 0)$	$(-\tau_2, 1, 0)$	$\left(\tau_1, 1, \frac{\kappa\omega^i}{3\delta}\right)$	$\left(-\tau_1, 1, \frac{\kappa\omega^i}{3\delta}\right)$
1	1	1	2	3
10	20	30	$1i$	$1i$
11	21	31	$2, i-1$	$2, i-2$
12	22	32	$3, i-2$	$3, i-1$

In the last two columns, i has the values 0, 1, 2; while $i - 1$ or $i - 2$ is to be replaced by the number 0, 1, 2 to which it is congruent modulo 3.

When F is the field of all real numbers, κ may be taken to be real, while just one of the numbers τ_1 and τ_2 is real. Hence 3 and only 3 of the 9 inflexion points are real. The same result is true if F is the field of the p residues of integers modulo p, where p is a prime $3j + 2 > 2$. For, κ may be taken to be integral (§ 4), while ω is imaginary and hence -3 is a quadratic non-residue of p. If $-t$ is a quadratic residue, τ_1 is real and τ_2 imaginary. If $-t$ is a non-residue, the reverse is true.

Next, let $p = 3j + 1$, so that ω is real and hence -3 a quadratic residue. By (5) and (6), τ_1 and τ_2 are both real or both imaginary according as $-t$ is a quadratic residue or non-residue of p. Hence all 9 inflexion points are real if and only if $-t$ is both a square and a cube and hence a 6th power modulo p. If $-t$ is a square but not a cube, only the first 3 inflexion points are real. If $-t$ is a quadratic non-residue, (1, 0, 0) is the only real inflexion point.

A cubic with integral coefficients taken modulo p, a prime > 3, with at least one real inflexion point and with invariant $s = 0$ and invariant $t \neq 0$, has 9 real inflexion points if $p = 3j + 1$ and $-t$ is a sixth power modulo p, a single real inflexion point if $p = 3j + 1$ and $-t$ is a quadratic non-residue of p, and exactly 3 real inflexion points in all of the remaining cases.

For example, if $p = 7$ and $s = 0$, $t \neq 0$, there are 9 real inflexion points only when $t \equiv -1$. Taking $\delta \equiv 3$, $\tau_1 \equiv -2$,

$\tau_2 \equiv +1$, $\kappa \equiv -1$, we get $\omega \equiv 2$. Thus $x^2y - y^3 + 3z^3 \equiv 0$ has the 9 inflexion points (1, 0, 0), (1, 1, 0), (− 1, 1, 0), $(-2, 1, 3 \cdot 2^i)$, $(2, 1, 3 \cdot 2^i)$ $(i = 0, 1, 2)$.

6. *Inflexion Points when* $s \neq 0$, $\Delta \neq 0$.—These are (1, 0, 0) and

$$(9) \qquad \left(\frac{s - k^2}{\pm 2k\sqrt{-k}}, \quad \frac{3\delta}{k}, \quad 1\right),$$

where k ranges over the roots of the quartic (3). We seek the number of real roots k for which $\sqrt{-k}$ is real. In order that the left member of (3) shall have the factors

$$(10) \qquad k^2 + wk + l, \quad k^2 - wk + m,$$

it is necessary and sufficient that

$$(11) \quad l + m - w^2 = -6s, \quad (l - m)w = t, \quad lm = -3s^2.$$

Let $t \neq 0$ (for $t = 0$ see § 9). Then $w \neq 0$ and

$$(12) \qquad 2l = w^2 - 6s + t/w, \quad 2m = w^2 - 6s - t/w.$$

Inserting these values into (11_3), we get

$$(13) \qquad w^6 - 12sw^4 + 48s^2w^2 - t^2 = 0.$$

Set $w^2 = y + 4s$. Then

$$(14) \qquad y^3 = t^2 - 64s^3 = 27\Delta.$$

For the rest of this section, let the field be that of the residues of integers modulo p, where p is an odd prime $3j + 2$. Since any integer e has a unique cube root e^{-j} modulo p, there is a single real root y of (14).

First, let $y + 4s$ be a quadratic residue of p. Then w is real and hence also l and m. The product of the discriminants of the quadratic functions (10) is seen by (11_1) and (11_3) to equal

$$(15) \qquad (w^2 - 4l)(w^2 - 4m) = -3(w^2 - 4s)^2 = -3y^2$$

and hence is a quadratic non-residue of p. Thus a single one of the quadratics (10), say the first, has a discriminant which is a quadratic residue and hence has real roots. By (12_1),

$$4l(w^2 - 4l)w^2 = -2w^6 - 6w^3t + 36sw^4 - 4t^2 + 48stw - 144s^2w^2.$$

Adding the vanishing quantity (13), we see that

$$(16) \qquad 4l(w^2 - 4l)w^2 = -3(w^3 - 8sw + t)^2.$$

Since $w^2 - 4l$ is a quadratic residue and -3 is a non-residue of p, it follows that l is a non-residue. Hence a single one of the roots of the first quadratic (10), and hence a single one of the roots of the quartic (3), is the negative of a quadratic residue. Thus just two of the inflexion points (9) are real.

Next, let $y + 4s$ be a quadratic non-residue of p. Then there is no factorization of the quartic (3) into real quadratic factors. Nor is there a real linear factor $k - r$ and a real irreducible cubic factor. For, if so, the roots of the latter are of the form $\lambda, \lambda^p, \lambda^{p^2}$ (cf. the first foot-note p. 37). Then

$$(r-\lambda)(r-\lambda^p)(r-\lambda^{p^2}),\ P=(\lambda-\lambda^p)(\lambda^p-\lambda^{p^2})(\lambda^{p^2}-\lambda)\equiv P^p \pmod{p}$$

are real, so that the discriminant of (3) is a quadratic residue. But this discriminant was seen to be $-3(81\Delta)^2$, and -3 is a non-residue. Hence (3) is irreducible modulo p. Thus (1, 0, 0) is the only real inflexion point.

For $p = 3j + 2 > 2$, a cubic (1) with $st\Delta \neq 0$, has exactly three real inflexion points or a single one according as the real number $3\Delta^{\frac{1}{3}} + 4s$ is a quadratic residue or non-residue of p.

7. *Cubic with $st\Delta \neq 0$, $p = 3j + 1$.*—Now -3 is a quadratic residue of p and there are three real cube roots $1, \omega, \omega^2$ of unity modulo p.

In this section we shall assume that Δ is a cube modulo p. Then there are three real roots y_i of (14). At least one of the $y_i + 4s$ is a quadratic residue of p since

$$\prod_{i=1}^{3} (y_i + 4s) = y_1^3 + 64s^3 = t^2.$$

If $y_1 + 4s$ is a quadratic residue, while $y_2 + 4s$ and $y_3 + 4s$

are non-residues, there is a single factorization of quartic (3) into real quadratics (10) and hence certainly not four real roots. The product (15) of the discriminants of the real quadratic factors is now a quadratic residue of p. If each were a residue, there would be four real roots. Hence each is a non-residue and there is no real root. *There is a single real inflexion point if* $p = 3j + 1$, $st\Delta \neq 0$, Δ *is a cube, and if the three values of* $3\Delta^{\frac{1}{3}} + 4s$ *are not all quadratic residues of* p.

Next, let each $y_i + 4s$ be a quadratic residue of p. Then there are three ways of factoring quartic (3) into real quadratics (10). But a root common to two distinct real quadratics is real. Hence all four roots are real. The discriminant of each quadratic (10) is therefore a quadratic residue of p. Hence, by (16), l is a quadratic residue of p; similarly for the constant term of each quadratic factor. Thus the negatives of the four roots are all quadratic residues or all non-residues.

To decide between these alternatives, we need the actual roots. In $w_i^2 = y_i + 4s$, let the signs of the w_i be chosen so that

$$k^2 - w_i k + m_i = 0 \qquad (i = 1, 2, 3)$$

have a common root. As in (12),

$$2m_i = w_i^2 - 6s - t/w_i.$$

For $e \neq 1$, we find by subtraction and cancellation of $w_1 - w_e$ that

$$2k = w_1 + w_e + t/(w_1 w_e).$$

Comparing the results for $e = 2$ and $e = 3$, we get

$$w_1 w_2 w_3 = t. \tag{17}$$

Hence* the roots of (3) are

$$\begin{aligned} &\tfrac{1}{2}(w_1 + w_2 + w_3), \quad \tfrac{1}{2}(w_1 - w_2 - w_3), \\ &\tfrac{1}{2}(- w_1 + w_2 - w_3), \quad \tfrac{1}{2}(- w_1 - w_2 + w_3). \end{aligned} \tag{18}$$

The product of the first and $(i + 1)$th roots is seen to equal m_i

* In particular, we have deduced Euler's solution by the method of Descartes.

and hence is a quadratic residue. For given values of p, s, t, we can readily find by a table of indices the real values of the w_i and thus a real root and hence decide whether or not it (and hence each of the four roots) is the negative of a quadratic residue.

However, changing our standpoint, we shall make an explicit determination of all sets s, t for which the quartic (3) has four real roots each the negative of a quadratic residue of p.

By the definition of the w_i^2, or direct from (13),

$$(19) \qquad \Sigma w_1^2 = 12s, \quad \Sigma w_1^2 w_2^2 = 48s^2, \quad w_1^2 w_2^2 w_3^2 = t^2.$$

Let ω be a fixed integral root of $\omega^2 + \omega + 1 \equiv 0 \pmod{p}$. Then

$$0 = (12s)^2 - 3(48s^2) = \Sigma w_1^4 - \Sigma w_1^2 w_2^2$$
$$= (w_1^2 + \omega w_2^2 + \omega^2 w_3^2)(w_1^2 + \omega^2 w_2^2 + \omega w_3^2).$$

Interchanging w_2 and w_3, if necessary, we have

$$(20) \qquad w_1^2 + \omega w_2^2 + \omega^2 w_3^2 \equiv 0 \qquad \pmod{p}.$$

Conversely, if the w_i^2 are any quadratic residues satisfying (20) and if we define s and t by (19_1) and (17), we obtain a quartic (3) with the four real roots (18). If we permute w_1, w_2, w_3 cyclically we obtain solutions of (20) leading to the same s and t and to the same four roots (18).

Our first problem is therefore to find all sets of solutions of (20). To this end it is necessary to treat separately the cases -1 a quadratic residue and -1 a non-residue; viz., $p = 12q + 1$ and $p = 12q + 7$ (since already $p = 3j + 1$).

First, let $p = 12q + 1$. Then $-1 \equiv i^2 \pmod{p}$, where i is an integer. Set

$$2\rho = w_1 - i\omega w_3, \quad 2\sigma = w_1 + i\omega w_3.$$

Then (20) becomes

$$4\rho\sigma = -\omega w_2^2 = (i\omega^2 w_2)^2,$$

so that $\rho\sigma$ must be a quadratic residue. Hence we may take

$\sigma = \rho l^2$, where ρ and l are integers not divisible by p. Then

(21) $\quad w_1 = \rho(1 + l^2), \quad w_2 = 2i\omega\rho l, \quad w_3 = i\omega^2\rho(1 - l^2).$

We must exclude the values of l which lead to equal values of two of the w_i^2, and hence to equal y_i's, since the roots of (14) are incongruent. Now if any two of the w_i^2 in (20) are congruent, all three are congruent. But $w_1^2 \equiv w_2^2$ implies

$$1 + l^2 \equiv \pm 2i\omega l, \quad (l \mp i\omega)^2 \equiv \omega^4, \quad l \equiv \pm i\omega + e\omega^2 \qquad (e^2 \equiv 1).$$

The values $l^2 \equiv 0, \pm 1$ make one of the $w_i \equiv 0$. Hence we must exclude the 9 incongruent integral values of l:

(22) $\quad l = 0, \quad \pm 1, \quad \pm i, \quad \omega^2 \pm i\omega, \quad -\omega^2 \pm i\omega.$

Using the values (21), we get

(23) $\quad 12s = \rho^2\{(1 - \omega)(1 + l^4) - 6\omega^2 l^2\}, \quad t = 2\rho^3 l(l^4 - 1),$

(24) $\quad \frac{1}{2}(w_1 + w_2 + w_3) = \frac{1}{2}\rho(1 + i\omega^2)\left(1 + \frac{i\omega l}{1 + i\omega^2}\right)^2.$

To make the negative of the last a square, we must take

(25) $\quad \rho = -2(1 + i\omega^2)r^2 \qquad (r \not\equiv 0).$

Now s, given by (23), is zero only when

(26) $\quad l = \omega \pm i\omega^2, \quad -\omega \pm i\omega^2.$

The desired sets s, t are given by (23) *and* (25), *where r is any integer not divisible by p, while l is any one of the $p - 13$ positive integers $< p$ not congruent modulo p to one of the* 13 *incongruent integers* (22), (26). *The minimum p is* 37.

Second, let $p = 12q + 7$. Then $\lambda^2 \equiv -1 \pmod p$ is irreducible. Its roots i and $-i = i^p$ are Galois imaginaries. Set

(27) $\quad \pi = p + 1, \quad \sigma = p - 1.$

There exists a linear function R of i with integral coefficients such that $R^{\pi\sigma} = 1$, while no lower power of R is unity. Any function of i is zero or a power of R and any integer is a power of

R^{π}, a primitive root of p. Hence we may set

$$\omega^2 w_2 = R^{\pi\eta}, \quad w_1 + \omega w_3 i = R^e, \quad w_1 - \omega w_3 i = R^{pe},$$

where $0 \leqq \eta < \sigma$, $0 \leqq e < \pi\sigma$. Then (20) is equivalent to

$$R^{\pi e} + R^{2\pi\eta} = 0, \quad \pi e \equiv 2\pi\eta + \tfrac{1}{2}\pi\sigma \qquad (\mathrm{mod}\ \pi\sigma).$$

The last condition is equivalent to

$$(28) \qquad e = 2\eta + \sigma/2 + j\sigma \qquad (0 \leqq j < \pi).$$

We have

$$w_2 = \omega R^{\pi\eta}, \quad 2w_1 = R^e + R^{pe}, \quad 2w_3 = -i\omega^2(R^e - R^{pe}),$$

$$2\omega^2 \Sigma w_1 = 2R^{\pi\eta} + (\omega^2 - i\omega)R^e + (\omega^2 + i\omega)R^{pe},$$

$$(29) \qquad \begin{gathered} (\omega^2 - i\omega)(\omega^2 + i\omega) = -1, \\ (\omega^2 - i\omega)^{\pi} = -1, \quad \omega^2 - i\omega = R^{f\sigma/2} \qquad (f \text{ odd}), \end{gathered}$$

$$\begin{aligned} 2\omega^2 \Sigma w_1 &= 2R^{\pi\eta} + R^{e+f\sigma/2} - R^{pe-f\sigma/2} \\ &= R^{\pi[j-(f+1)/2]}(R^{\eta-j+p(f+1)/2} + R^{p\eta-pj+(f+1)/2})^2. \end{aligned}$$

The last binomial is its own pth power and hence is real. We desire that the root $\frac{1}{2}\Sigma w_1$ shall be the negative of a quadratic residue and hence a non-residue. Since R^{π} is a primitive root of p, the condition is that $j - (f + 1)/2$ shall be odd:

$$(30) \qquad f = 2l - 1, \quad j - l = \text{odd}.$$

We must exclude the values making $w_1^2 = w_2^2$:

$$0 = 2R^{\sigma/2}(w_1 \mp w_2) = R^{2\eta+\sigma+j\sigma} \mp 2\omega R^{\pi\eta+\sigma/2} - R^{2p\eta-j\sigma},$$

the second term having been simplified by use of

$$R^{\pi\sigma/2} = -1, \quad R^{p\sigma} = R^{-\sigma}.$$

Completing the square of the first two terms, we get

$$(R^{\eta+\sigma(j+1)/2} \mp \omega R^{p\eta-\sigma j/2})^2 = (\omega^2 + 1)R^{2p\eta-\sigma j}.$$

Now $\omega^2 + 1 = -\omega = (ci\omega^2)^2$, where $c = 1$ or -1. Hence

$$R^{\eta+\sigma(j+1)/2} = (\pm\omega + ci\omega^2)R^{p\eta-\sigma j/2}.$$

But

$$(31)\qquad \begin{aligned}(\omega + i\omega^2)(\omega - i\omega^2) &= -1, \quad \omega + i\omega^2 = R^{v\sigma/2},\\ \omega - i\omega^2 &= -R^{-v\sigma/2} \qquad (v \text{ odd}).\end{aligned}$$

Hence we must exclude the four cases in which

$$(32)\qquad \eta \equiv j + \tfrac{1}{2}(\pm v + 1), \quad j + \tfrac{1}{2}(\pm v + \pi + 1) \qquad (\text{mod } \pi),$$

these four values being incongruent.

No one of the w's in (29) is zero, since e is odd by (28), so that $e \not\equiv 0, \pi/2 \pmod{\pi}$. By (19_1) and (17),

$$(33)\qquad \begin{aligned}48s &= (1 - \omega)(R^{2e} + R^{2pe}) + 6\omega^2 R^{2\pi\eta},\\ 4t &= -iR^{\pi\eta}(R^{2e} - R^{2pe}).\end{aligned}$$

Finally, we must here exclude the cases in which $s = 0$. Combining $\Sigma w_1^2 = 0$ with (20), we obtain the necessary and sufficient condition $w_1^2 = \omega w_3^2$ for $s = 0$. But $w_1 = \pm\omega^2 w_3$, in connection with (29), gives

$$R^e(1 \pm i\omega) = R^{pe}(-1 \pm i\omega), \quad R^e(\omega \pm i\omega^2)^2 = R^{pe}.$$

Thus, by (31), the condition is that $e \pm v\sigma \equiv pe \pmod{\pi\sigma}$ or $e \equiv \pm v \pmod{\pi}$. Then, by (28), η is congruent modulo π to one of the values (32) decreased by $\pi/4$. *Hence the desired sets s, t are given by* (33), *subject to* (28), *in which the* 8 *incongruent η's given by* (32) *and those values decreased by $\pi/4$ are excluded. In particular, $p > 7$.*

For $p = 19$, the only admissible pairs are

$$s = 2 \cdot 2^{2l}, \quad t = 6(-2)^{3l} \quad (l = 0, 1, \cdots, 8).$$

For any l, the negatives of the roots of quartic (3) are the products of $-3 \equiv 4^2, 4, 7 \equiv 8^2, -8 \equiv 7^2$ by $(-2)^l$ and hence are quadratic residues of 19 since $-2 \equiv 6^2$.

For $p = 31$, the only pairs are

$$s = 3^{2l}, \quad t = 5(-3)^{3l}; \quad s = -3^{2l}, \quad t = 13(-3)^{3l} \quad (l = 0, \cdots, 15),$$

the negatives of the roots of (3) being the products of $7, -11, -12, -15$ and $-3, 5, 9, -11$, respectively, by $(-3)^l$, and hence are quadratic residues of 31.

8. *Case* $p = 3j + 1$, $st\Delta \neq 0$, Δ *not a Cube.*—The roots of (14) are now Galois imaginaries y, y^p, y^{p^2}. As at the beginning of § 7,

$$t^2 = (y + 4s)(y^p + 4s)(y^{p^2} + 4s) \equiv (y + 4s)^{1+p+p^2}.$$

Raise each member to the power $(p - 1)/2$. We see that $y + 4s$ is the square of an element, say w, of the Galois field of order p^3. The first root (18) is $\frac{1}{2}(w + w^p + w^{p^2})$ and equals its own pth power, and hence is real. This is not true of the remaining roots (18), since $w^p \neq w$, or since a real quadratic factor would imply that w is real. Hence *the quartic has a single real root.*

For $p = 7$, the only cases in which the negative of the single real root is a quadratic residue are $t = -1$ or 3, $s = -1, -2, 3$; $t = 2$, s arbitrary $\neq 0$. For $p = 13$, the only cases are

$$\pm t = 4, 5, 6; \quad s = -1, -3, 4 \qquad (s^3 \equiv -1);$$

$$\pm t = 1, 5, 6; \quad s = -2, -5, -6 \qquad (s^3 \equiv 5);$$

and $\pm t = 3$, $-s$ equals one of the preceding six values of s.

9. *Cubic with* $t = 0$, $s \neq 0$.—In this case, (3) becomes

$$(k^2 - 3s)^2 = 12s^2.$$

If there be a real root k, 3 is a quadratic residue of p, and

$$k^2 = ls, \quad l = 3 \pm 2\sqrt{3}.$$

First, let $p = 3j + 2$, so that -3 is a quadratic non-residue of p. Then -1 must be a non-residue of p and hence $p = 12r + 11$. The product of the two l's is -3, so that a single value of k^2 is a quadratic residue. Since the two real k's are of opposite sign, there is a single real root k whose negative is a quadratic residue. *For* $t = 0$, $s \neq 0$, *and* $p = 12r + 5$, *there is a single real inflexion point; for* $p = 12r + 11$, *there are just three real inflexion points.*

Finally, let $p = 3j + 1$, so that -3 is a quadratic residue of p. If $p = 12r + 7$, then 3 is a non-residue, and there is no real k and hence a single real inflexion point. If $p = 12r + 1$, the four roots k are all real or all imaginary. For $p = 13$, $k^2 \equiv -2s$ or $-5s$, and $-k$ is a quadratic residue if and only if $k^6 \equiv 1$, $s^3 \equiv 8$, $s \equiv 2, 5, 6$. For $p = 37$, $k^2 \equiv -4s$ or $10s$, and $-k$ is a residue if and only if $s^9 \equiv 1$.

The end

A series of hardcover reprints of major works in mathematics, science and engineering.
All editions are $5\frac{5}{8} \times 8\frac{1}{2}$ unless otherwise noted.

Mathematics

Theory of Approximation, N. I Achieser. Unabridged republication of the 1956 edition. 320pp. 49543-4

The Origins of the Infinitesimal Calculus, Margaret E. Baron. Unabridged republication of the 1969 edition. 320pp. 49544-2

A Treatise on the Calculus of Finite Differences, George Boole. Unabridged republication of the 2nd and last revised edition. 352pp. 49523-X

Space and Time, Emile Borel. Unabridged republication of the 1926 edition. 15 figures. 256pp. 49545-0

An Elementary Treatise on Fourier's Series, William Elwood Byerly. Unabridged republication of the 1893 edition. 304pp. 49546-9

Substance and Function & Einstein's Theory of Relativity, Ernst Cassirer. Unabridged republication of the 1923 double volume. 480pp. 49547-7

A History of Geometrical Methods, Julian Lowell Coolidge. Unabridged republication of the 1940 first edition. 13 figures. 480pp. 49524-8

Linear Groups with an Exposition of Galois Field Theory, Leonard Eugene Dickson. Unabridged republication of the 1901 edition. 336pp. 49548-5

Continuous Groups of Transformations, Luther Pfahler Eisenhart. Unabridged republication of the 1933 first edition. 320pp. 49525-6

Transcendental and Algebraic Numbers, A. O. Gelfond. Unabridged republication of the 1960 edition. 208pp. 49526-4

Lectures on Cauchy's Problem in Linear Partial Differential Equations, Jacques Hadamard. Unabridged reprint of the 1923 edition. 320pp. 49549-3

The Theory of Branching Processes, Theodore E. Harris. Unabridged, corrected republication of the 1963 edition. xiv+230pp. 49508-6

The Continuum, Edward V. Huntington. Unabridged republication of the 1917 edition. 4 figures. 96pp. 49550-7

Lectures on Ordinary Differential Equations, Witold Hurewicz. Unabridged republication of the 1958 edition. xvii+122pp. 49510-8

Mathematical Methods and Theory in Games, Programming, and Economics: Two Volumes Bound as One, Samuel Karlin. Unabridged republication of the 1959 edition. 848pp. 49527-2

Famous Problems of Elementary Geometry, Felix Klein. Unabridged reprint of the 1930 second edition, revised and enlarged. 112pp. 49551-5

Lectures on the Icosahedron, Felix Klein. Unabridged republication of the 2nd revised edition, 1913. 304pp. 49528-0

On Riemann's Theory of Algebraic Functions, Felix Klein. Unabridged republication of the 1893 edition. 43 figures. 96pp. 49552-3

A Treatise on the Theory of Determinants, Thomas Muir. Unabridged republication of the revised 1933 edition. 784pp. 49553-1

A Survey of Minimal Surfaces, Robert Osserman. Corrected and enlarged republication of the work first published in 1969. 224pp. 49514-0

The Variational Theory of Geodesics, M. M. Postnikov. Unabridged republication of the 1967 edition. 208pp. 49529-9